This Old House Sourcebook

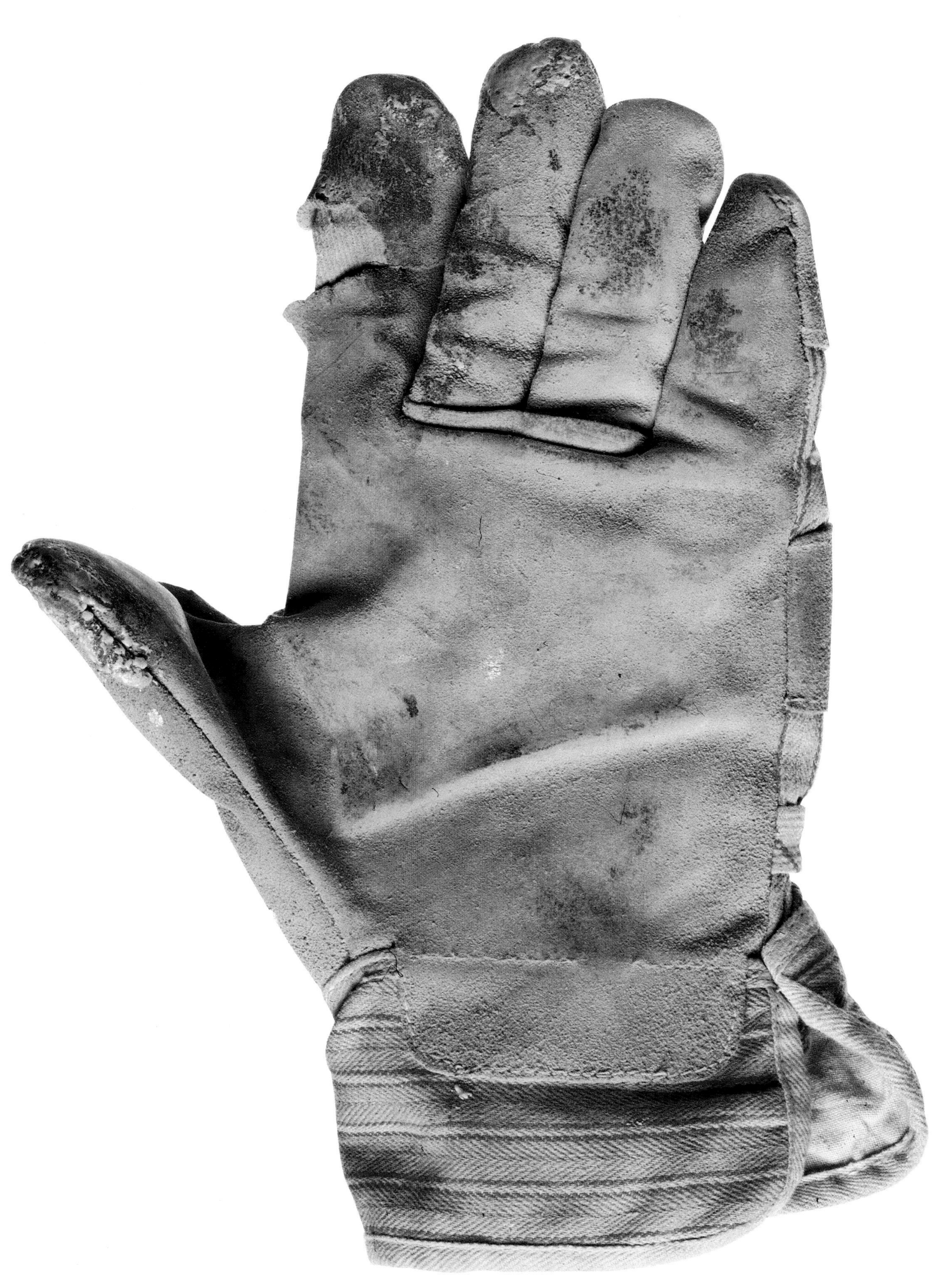

This Old House Sourcebook

Where to Find and How to Use Tools and Materials to Fix and Improve Your Home

The Editors of This Old House Magazine

This Old House Books

First Edition

ISBN 0-316-83171-9

Library of Congress Catalog Card Number 96-80503

10 9 8 7 6 5 4 3 2 1

Published simultaneously by Little, Brown and Company

Printed in the United States of America

Project Editor: Laura Goldstein
Editor: Bruce Shostak
Copy Editor: Mila Drumke
Contributing Editor: Pamela Hartford

Design Director: Matthew Drace
Art Director: Lisa Wagner
Designer: Michael Holtermann
Production Director: Denise Clappi

Front Cover Photograph of Norm Abram and Steve Thomas: David Katzenstein
Back Cover Photograph of Norm Abram, Steve Thomas, Tom Silva and Richard Trethewey: Dan Borris
Cover Illustration and Chapter Opener Illustrations: Clancy Gibson

foreword

"Where can I get this?" is the question most frequently asked of the staff of *This Old House* magazine. The specific *this* may be a commercial product used in home renovation or maintenance, the address of a supplier of specialty drywall or recycled glass tiles, the manufacturer of a hand or power tool, or a description of a building technique or process.

It was inevitable that a magazine in the extremely popular field of home improvement should become a clearinghouse for information not readily available in a neighborhood home center, a hardware store or the Yellow Pages. The quantity of inquiries we get is a vote of confidence in the quality and relevance of information we research, double-check and then publish in every issue.

This volume gathers together source information for tools and materials used in the magazine since its premiere issue in May 1995. It is not a comprehensive reference for home renovation and maintenance; we only cite items we have actually used. As the magazine continues to grow, we will revise and expand our sourcebook and publish similar books on individual topics such as landscaping, gardening and other outdoor projects. As we learn more, so will you.

—The editors of *This Old House* magazine

workshop

wood products

roofing, siding and windows

contents

walls and ceilings

kitchens and baths

fasteners

salvaged and recycled materials

introduction

When *This Old House* began renovating homes for television audiences, Norm Abram was the general contractor and did a lot of the construction work himself. He also did most of the buying of materials and supplies, usually purchasing them at several of the large professional supply houses that served the Boston area. But invariably, he would have to make a quick run to the hardware store to get a few things needed on site each day. Director Russ Morash would often go with him—filming had to wait until they got the right drill bit or the longer bolts. Inevitably, they would end up at Spags's place.

Spags (short for his boyhood nickname of Spaghetti—his real name was Anthony Borgatti) had been in the business for years, acquiring a vast mental inventory of every washer and clamp the store ever carried. There never came a need to organize the merchandise since it could all be located in Spags's head. If you couldn't find something, you asked him. He was quite opinionated, always happy to let you know which brand of screwdriver would never last the whole day, and he would freely convey his suspicions about the usefulness of any new special-purpose adhesive. The crew was always grateful for this information, especially because its truthfulness was usually borne out by subsequent experience.

Spags's store was not exceptionally lucrative, because he could only sell as many things as he had time to talk about, but he compensated for this with several ingenious cost-saving measures. He purchased a few old trailers and arranged them around his ramshackle site. He made deals with all kinds of suppliers. "What's your lowest price?" he'd demand. They would hem and haw and say, "Well, how many do you want?" "Just tell me your lowest price!" They'd offer a price, and Spags would respond, "Send me a truckload!" Eventually, the trailers were piled with merchandise at rock-bottom prices. Spags relished the dealmaking and was often tempted by items not quite related to the hardware business, such as the truckload of Lifesavers candy he laid in one summer. Shopping at the store took on a flea-market quality. Everyone always walked away with more than they'd planned to buy.

Spags's other cost-saving trick was not to provide customers with paper or plastic bags to carry the merchandise out of the store. He wasn't in the packaging business, he would remind people, he was in the hardware business. If you wanted to take something out of his store, you had to bring your own bag—or buy one from him. The only exception he made was for nails. If you wanted less than a 10-pound box, he would weigh out the nails with great exactitude and put them in the

GP
NOLAN
JNC

smallest, thinnest paper bag he could find. It was always best to make sure you had your own bags to dump the nails into. Spags's thin bags would split as soon as a harried carpenter tried to scoop out a handful. That's why Norm and the crew dubbed him "Spags No-Bags."

After Spags died last year, his daughters took over the store. But, like everyone else, *This Old House* has moved on to do most of its shopping at home centers, chain hardware stores and renovation supply houses. It's hard not to miss the personal, quirky nature of shopping at Spags No-Bags. It was reassuring. Today there are fewer and fewer small, family-run businesses, the kind where you can ask questions and know you'll get an answer from someone with long years of experience, the kind of place where the floorboards creak and the prices are scrawled in pencil, not laser printed.

The other side of this story is that the tools and materials available to do-it-yourselfers have greatly improved in quality. The volume commanded by large home-center stores makes them a powerful force for manufacturers, who have responded by making higher quality, more professional tools for the do-it-yourselfer. Homeowners, more involved than ever in decisions about insulation, siding, maintenance and repair, have spurred manufacturers to come up with more and more products.

Even with all the improved tools and materials out there, we're the first to acknowledge there are plenty of moments of frustration in home renovation and repair. The first and most important step in tackling a project, both in terms of time and dollars, is to be completely frank with yourself about how much time it will take and whether or not you possess the skills to do the job. Show host Steve Thomas often asks these questions of homeowners when they offer to undertake certain tasks in a renovation. He advises them to make a plan for the steps of the project, to figure out what materials and tools are needed and what they will cost, and then to analyze whether it's worth it to do it themselves, hire someone to do it or combine both approaches. Some parts of the job, such as demolition, shopping for materials and finish work, are more easily handled than others. If you have the time and want to learn a skill associated with the project, like soldering pipe or applying veneer plaster, then build the time and cost of learning into the budget.

With all the different places to shop for home renovation materials these days, it's a job just to figure out where the best buy is, where you can get the most reliable information or who is most likely to have what you need. We advise shopping around, first by telephone, to get prices, especially of large-ticket purchases. Home centers have great deals on some items, specialty suppliers or hardware stores are better for others.

All of us at *This Old House* magazine, with the collective experience of many projects behind us, have tried a lot of products available on the market. Manufacturers send us samples, and the various tradespeople we work with on each issue introduce us to new products and methods all the time. We're always on the lookout for high-quality and valuable innovations. We hope you'll use this sourcebook with confidence, making it one of the tools that helps you through doing it yourself.

SKILSAW

notes

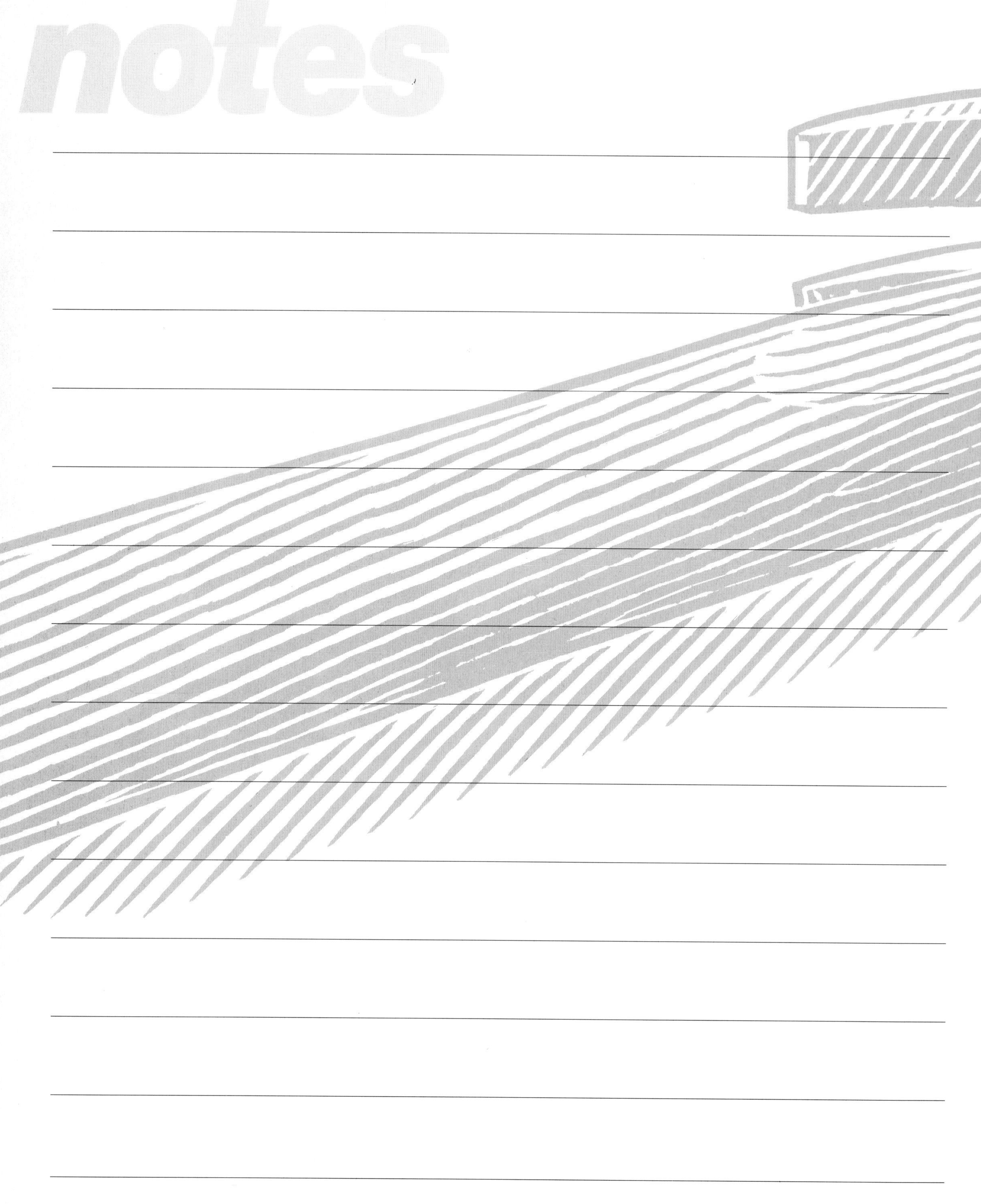

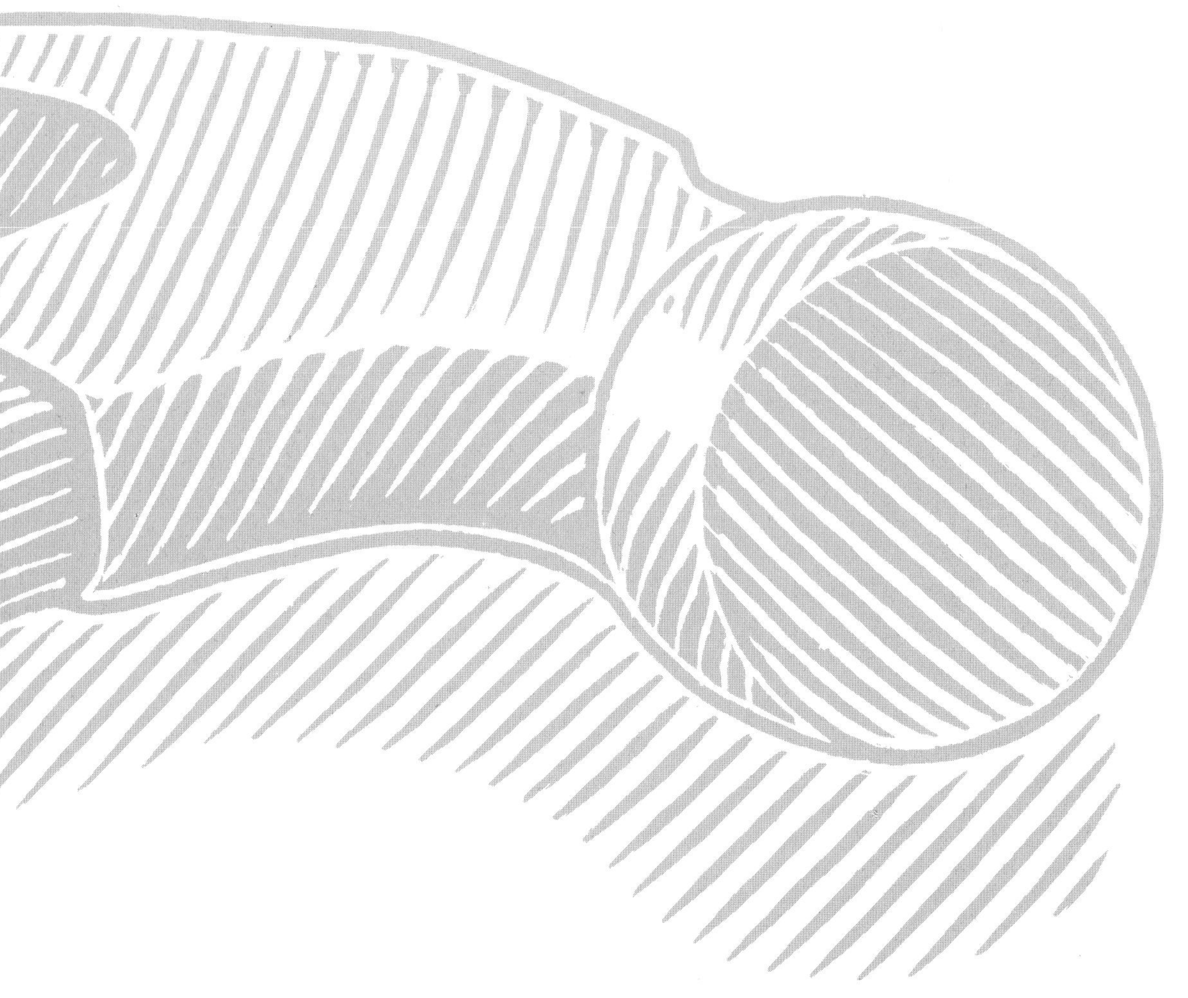

workshop

ABRASIVES | CLAMPS | CLEANING AIDS

CORDLESS DRILL/DRIVERS | HAMMERS | MEASURING TOOLS

PLANES | ROUTERS AND CARVERS | POWER SANDERS | HANDSAWS

POWER SAWS | SCREWDRIVERS | SHARPENING | SOLDERING

TOOL BELTS | UTILITY KNIVES | WORKBENCHES | WRENCHES AND PLIERS

abrasives

SANDPAPER

A century after the first synthetic abrasives were invented, garnet sandpaper still figures prominently in many hardware store displays. We reach for it because it's what our fathers used, and because it works. But garnet wears down faster than synthetic abrasives such as aluminum oxide and silicon carbide, so it's never used in the more highly engineered sandpapers sought out by professionals. These sandpapers don't work better just because they have different abrasives: The type of backing, the kind of adhesive that holds the abrasive to it and the frequent presence of special coatings are at least as important.

For example, there's a new light-green paper from 3M that minimizes clogging even on water-based finishes and polyurethanes. Its secret is not its abrasive (aluminum oxide, invented in 1897) but a soaplike coating that sloughs off, carrying sanding debris with it. A different 3M aluminum-oxide product is known as micron paper, after the dimension used to measure grit size. First developed for the computer industry, micron papers have a waterproof polyester film base, used because the superfine particles needed to polish parts would sink into paper or cloth. Grit size is closely controlled, so virtually all grains are the same size, compared with up to 50 percent filler particles in standard sandpaper. The better backing and more uniform sizing can give you the finish you want faster.

But micron papers are hard to find. 3M sells its Imperial Microfinishing Film to the furniture industry, where manufacturers willingly pay more for a product that delivers fast, consistent results even if workers lack experience. Sounds like a perfect product for do-it-yourselfers, right? Curiously, 3M doesn't market the paper to home-oriented stores because it believes consumers shop by price alone. A different strategy explains why another superior sandpaper—made with a 3M synthetic mineral called ceramic aluminum oxide—is marketed to professional floor finishers but not to companies that rent floor sanders to do-it-yourselfers. Rental stores want sandpaper that wears out fast so they can sell lots of it, says Galen Fitzel, 3M's technical service specialist for floor sanding products.

HOW TO READ SANDPAPER

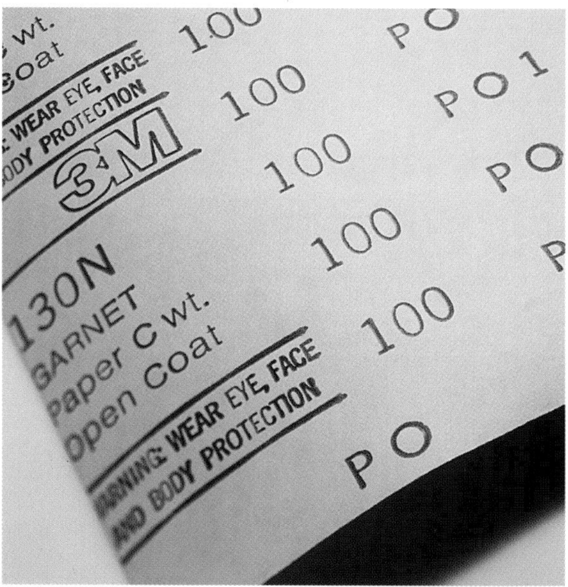

Safety Warning: Yes, a disc can fly off a random-orbit sander or a belt can break.

Product Number: Here it's 130N. This allows a user to reorder the exact sandpaper or call the manufacturer to inquire about its characteristics. These numbers are scattered around to make sure at least one label remains on each item when 4-foot-wide rolls are cut into belts, discs and sheets.

Abrasive: Garnet and other natural abrasives are usually listed, but synthetics are often identified just by trademarked names. For translations, call the manufacturer. At 3M, Three-M-ite and Production are aluminum oxide; Imperial can be aluminum oxide or silicon carbide; Tri-M-ite is silicon carbide; and Cubitron, Regal and Regalite are ceramic aluminum oxide.

Backing: Most hand sanding sheets are backed with paper. "A" weight, the lightest, is for hand sanding with fine grits. "C" and "D" weights are for hand and light power sanding of wood and drywall. "E" and "F" weights are used mainly in drum and belt sanders. Other backing options include cloth (most common for home use is "J" weight, also called jeans cloth), fiber and polyester film.

Open Coat: Sometimes abbreviated OC, this means abrasive particles cover 40 to 70 percent of the surface. The other option is closed coat, which means the surface is completely covered. Open-coat papers clog less quickly but leave a rougher surface.

Grit Size: 100 grit, like this example, means particles passed through a screen with 100 openings per inch (10,000 per square inch) but were trapped by a finer screen one grade up, with 120 openings. A "P" before the number signals a more tightly regulated European standard. A U-like micron symbol before a grit number means the least size variation. Micron No. 60 is equivalent to standard 220- to 240-grit paper.

Lot Number: The PO1 would allow the manufacturer to trace this sheet to the time and conditions under which it was manufactured, useful when problems develop. Belts have a second lot number, in a different ink, to identify the splice.

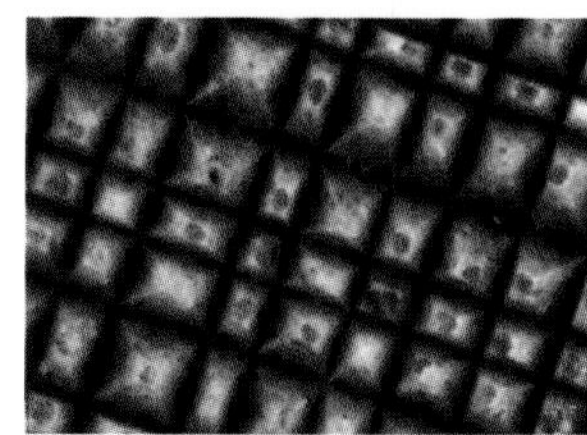

The newest, most revolutionary sandpaper is made by a process called microreplication, which molds the abrasive into a uniform pyramid pattern that makes consistent scratches without gouging.

SOURCES

Green sandpaper for water-based finishes and polyurethanes: Professional Painters' Abrasive, #235U, 80 cents to $1.05, depending on grit size, for 9"x11" sheet
3M Construction Markets Division, 3M Center, Building 225-4S-08, St. Paul, MN 55144-1000; 800-480-1704.

Micron sandpaper: Imperial Microfinishing Film, #258L, 100-micron grit, $27 for box of fifty 5" discs
R.S. Hughes Co., 1162 Sonora Court, Sunnyvale, CA 94086; 408-739-3211; manufactured by 3M Abrasive Systems Division, Building 223-6N-01, St. Paul, MN 55144; 800-742-9546.

Microreplication sandpaper: 3M Apex, #207EA, 180 grit, $239.75 for box of 25 2"x132" belts
Industrial Tool & Supply, 1177 N. 15th St., San Jose, CA 95112; 800-366-4281 or 408-292-8853; manufactured by 3M Abrasive Systems Division; 800-742-9546.

THE CHOICE OF ABRASIVES

Virtually all abrasives are crystals that break down during use, so sharp, fresh surfaces are repeatedly exposed. Abrasives are rated on two qualities—toughness (how hard it is to get the crystals to break) and hardness (how resistant they are to wearing down). Examples are magnified to show detail.

NATURAL

1. **GARNET: Softest and least tough of all common abrasives. Crystals fracture under light pressure, so garnet works well for hand sanding of wood. It's probably the best option for difficult power sanding jobs where other papers might burn the wood, such as sanding end grain of hardwood. Garnet is unsuitable for use on metal.**

2. **EMERY: Good for polishing metal but unsuitable for use on wood because the edges tend to dull rather than chip. The second hardest natural abrasive after diamond. Emery, a combination of corundum (a natural aluminum oxide) and iron oxide, is usually used with an oil lubricant.**

3. **CROCUS: An iron oxide, natural or synthetic, with a small amount of silicon dioxide. Like emery, crocus is used to clean and polish metal. But crocus is considerably softer than emery, so it is used where only very slight stock removal is desired. The same abrasive is the basis of rouge, often used to polish or buff metal.**

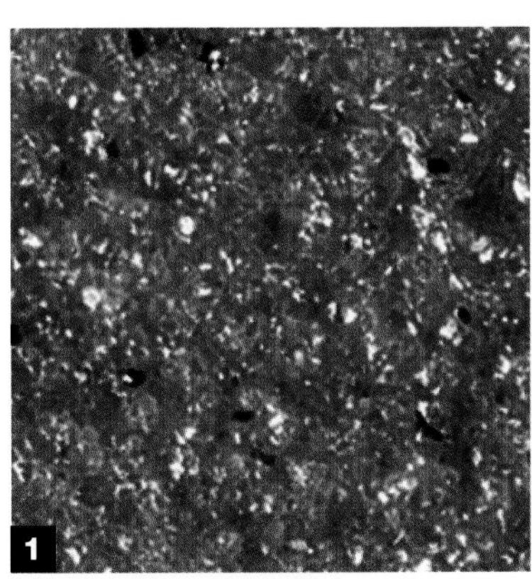
1

2

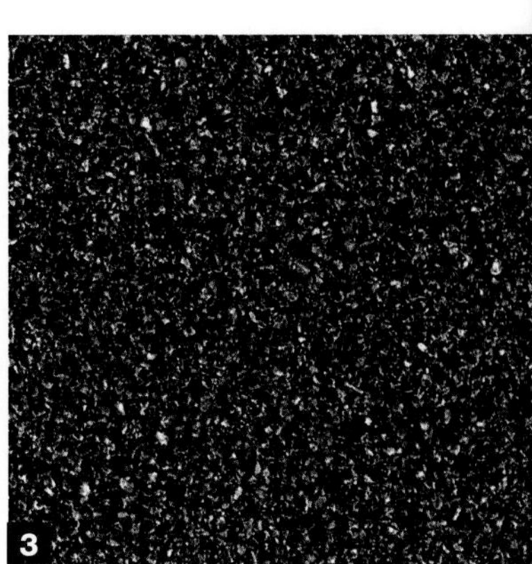
3

SYNTHETIC

4. **ALUMINUM OXIDE: The most common general-purpose abrasive for wood and metal. First made about 1897 from bauxite. The formula can be varied to produce crystals with very different characteristics. Here they are wedge-shaped (and embedded in a thick resin) to stand up well for floor sanding.**

5. **SILICON CARBIDE: The second most common abrasive, it is harder than aluminum oxide and produces a smoother surface on wood. It's the best choice for sanding between tough finish coats. Lubricate fine grits with water or oil to keep the sharp slivers from clogging up before they wear down. Also used on nonferrous metals.**

6. **CERAMIC ALUMINUM OXIDE: A new kind of aluminum oxide, two to three times tougher than the standard kind and with a more uniform crystalline structure. Good for power sanding of metal, wood and many other materials. Sought out by professional floor sanders. Often mixed with regular aluminum oxide for belt sanding.**

7. **ALUMINA ZIRCONIA: Blocky, sharp crystals remove a lot of material fast. Good for shaping and grinding wood and metal, but not for polishing. Often sold in wide "planer" belts that sand boards to a desired thickness. Made from oxides of aluminum and zirconium.**

8. **DIAMOND: The hardest abrasive, excellent for shaping and smoothing metal, glass and other hard materials. Pointless to use on soft materials such as wood. Made with synthetic diamonds to reduce cost (but it's still about $2 a square inch). The dot pattern shown here minimizes clogging by allowing plenty of room for sanding debris.**

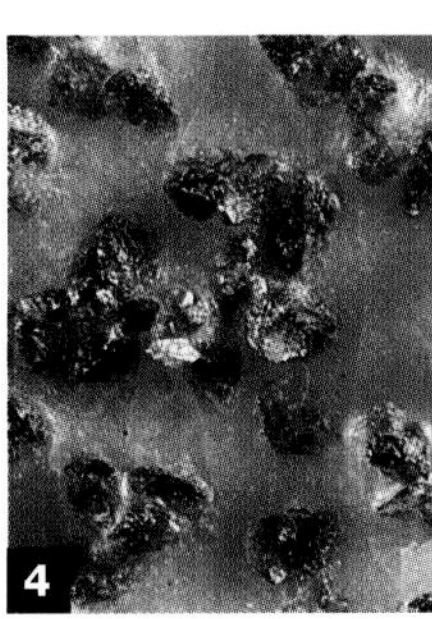
4

5

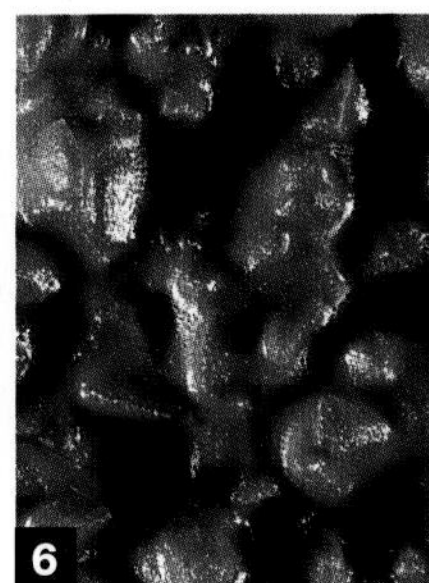
6

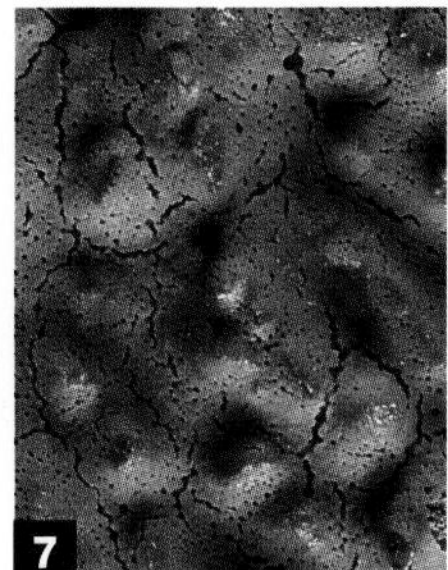
7

8

THE OTHER PARTS OF SANDPAPER

The backing: The stiffer it is, the faster the sandpaper will cut. Select a product that is no more flexible than it needs to be to conform to the shape being sanded. To increase the effective stiffness of thin sandpaper, use it with a rigid sanding block or pad.

The adhesive: Sandpaper has two layers of adhesive, one to anchor the grains and a second to lock them in place. The choices are two coats of hide glue, two coats of resin or a coat of resin over hide glue. Hide glue generally sets faster and is more flexible, but resin resists moisture and heat better. Paper with hide glue works well for sanding complex shapes, but moisture—even high humidity—can make it virtually useless. For power sanding and sanding with a lubricant, go for resin. Resin over hide glue produces sandpaper that starts out and stays sharpest the longest. The adhesive type is rarely specified, so one test is to exhale close to the surface and then sniff; if it stinks, the sandpaper contains hide glue.

REUSABLE ABRASIVES

Sandpaper is a great tool, but when it clogs up, it's darn near useless, even if the abrasive is still sharp. Here is a range of paperless sanding sheets that can be cleaned and reused again and again. The sanding gel is like grit in a bottle; just add more when you need it.

1. METAL SANDING SHEET
Tough enough to strip paint—and severely gouge wood.

2. FOAM SANDING BLOCKS
36 to 120 grit
For curved surfaces; two grits per block; rinses clean.

3. FOAM-BACKED SANDING SHEETS
60 to 150 grit
Perfect for wet sanding; residue washes away.

4. FOAM-BACKED SANDING PAD
60 to 220 grit
Thick foam for easy hand sanding, wet or dry.

5. SANDING GEL ON FOAM APPLICATOR PAD
3,000 to 10,000 grit
Polishes any nonporous surface.

SOURCES

Metal sanding sheets: Dragonskin #3331, 4½"x5", $2.99; with holder, #3309, $4.98
Red Devil Inc., 2400 Vauxhall Rd., Union, NJ 07083; 800-423-3845.

Foam sanding blocks: #907 (extra fine/fine), #908 (fine/medium), #909 (medium/coarse), $2.09 each
3M Product Information Center, Building 515-3N-06, St. Paul, MN 55144; 800-364-3577.

Foam-backed Rinse and Reuse sanding sheets: #CS-150 (fine), #CS-100 (medium), #CS-060 (coarse), two 9"x10" sheets, $3.99
NicSand Inc., Box 29480, Cleveland, OH 44129; 216-351-3333.

Foam-backed Contour sanding pad, twenty-four 4½"x5½" sheets, $1.56
3M Product Information Center.

Sanding gel and pad: #NSG-111, $9.99
NicSand Inc.

Sandflex blocks, 3"x2"x¾", $4
Klingspor Corp., Box 5069, Hickory, NC 28603-5069; 704-326-9663.

RUST ERASER

This sanding block erases corrosion fast. Unlike foam pads with an abrasive coating, Sandflex is rubber impregnated with an abrasive. It's long-lasting and can be carved to fit rounded railings. Dribble oil or water onto the block for best results.

clamps

PIPE CLAMPS

Pipe clamps have two components: the clamp fixtures, including the tail stop and head assembly, and the pipe. The tail stop slides in from one direction while the screw (part of the head assembly) applies pressure from the other. Some brands have a reversible head and a specially designed crank handle so the clamp can double as a spreader—handy for straightening bowed studs, for example. A spring-loaded clutch mechanism grabs the pipe and prevents the tail stop (and head assembly, if it's movable) from slipping around. This is the part of the fixture most likely to wear out. It's not worth replacing the mechanism, though parts are available from some manufacturers; it's easier and not much more expensive to simply pick up new clamp fixtures.

Fixtures are designed to fit ½- or ¾-inch pipe. Inexpensive steel "black" pipe is preferred by some. The surface is less slippery than that of galvanized pipe, so the clutch mechanism can dig in. But most contractors, including Tom Silva, find galvanized works just as well. Clamp fixtures and pipe are purchased separately at hardware stores or through tool suppliers. Tom picks up pipe free when he's demolishing old heating and plumbing systems.

EXTENDING A PIPE CLAMP'S REACH

One of the beauties of pipe clamps is that they're easy to extend. There's no need to haul around heavy, cumbersome lengths of pipe when it's possible to use a coupling—a pipe fitting with female threads at each end—to join two or more pipes for the reach needed to clamp together an entire run of cabinetry or glue up a large piece of furniture. To extend a pipe it must be threaded at both ends, one end to accept the head assembly and the other end for the coupling. Prethreaded pipe is available from lumber or plumbing supply stores. Salvaged pipe like Tom uses can be threaded at a plumbing supply store for a couple of dollars. Or, if you have a pipe die and plenty of extra time, you can thread it yourself.

PIPE CLAMPS AT WORK

Far left: Tom uses pipe clamps to hold tongue-and-groove V-boards in line while he joins them with battens. To prevent the jaws from marring the work, some clamps come with nonslip plastic pads. When they aren't available, Tom uses spare bits of cardboard or slices of wood, which he attaches to the jaws with double-stick tape.

Middle: Because pipe clamps are round, the head and tail can be rotated at different angles to accommodate disparate clamping surfaces. Here, the upper clamp holds three cabinets of differing heights in line.

Right: Tom is careful not to overtighten pipe clamps. Doing so forces too much glue out of the joints and dents the wood. Here he clamps together 1x4s to make a shelf for a microwave.

CLAMPS FOR EVERY JOB

Pipe clamps are great for big jobs, but you wouldn't use them to squeeze together the mitered corners on a picture frame. There are hundreds of different clamps, each suited to a different type of work. Most are versions of those shown here. When it comes to buying clamps, there are two rules to follow: Always buy them in pairs, and go for quality—there's nothing worse than having a clamp give way mid-project.

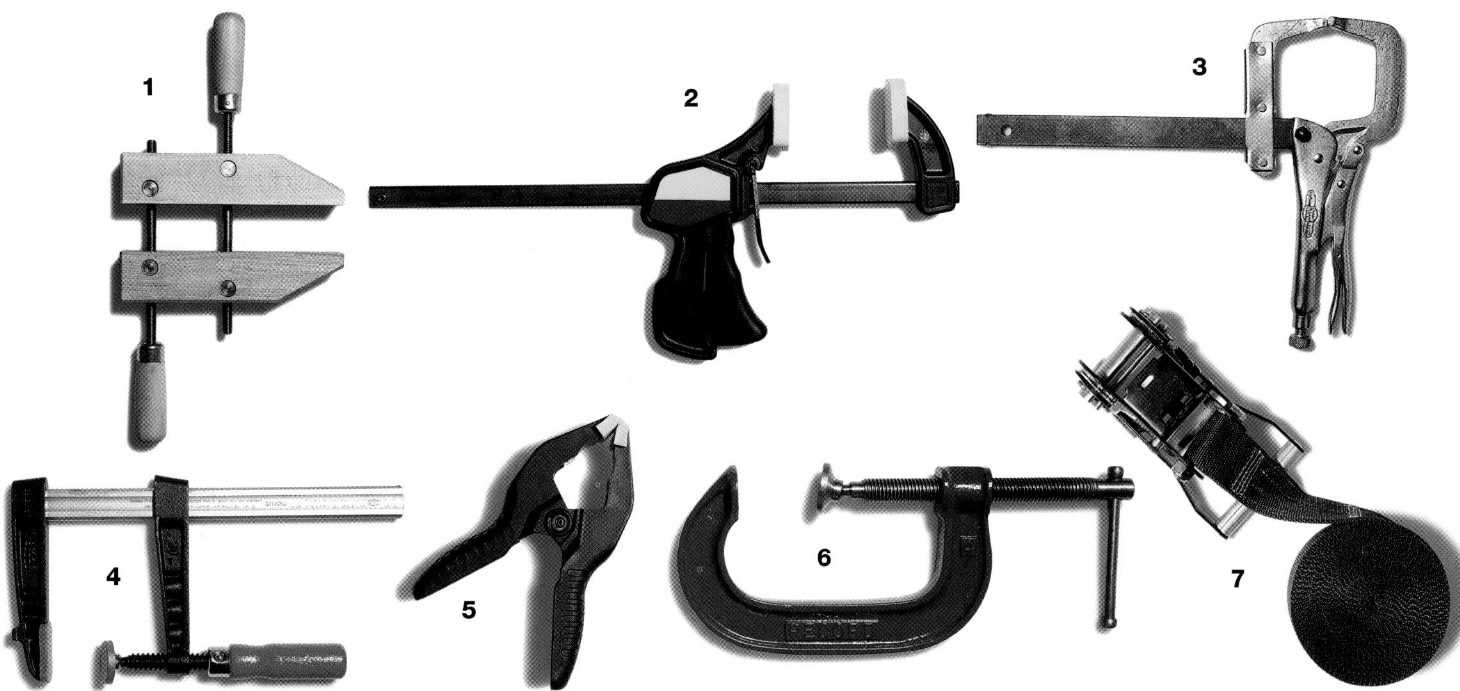

1. Wood hand-screw: Wide jaws spread pressure over a broad area. Dual action allows clamping at awkward angles. **2. One-handed bar:** These have done for carpenters what automatic bread slicers did for bakers. "One-handers operate so quickly and smoothly, I rarely reach for anything else," Tom says. The sliding head is tightened and adjusted with the pistol grip. The clamp releases immediately when the trigger is squeezed. **3. Locking bar:** Jaws can be set to a fixed depth, making these good for repetitious work. **4. Quick-action bar:** Deep jaws are handy for clamping hard-to-reach areas or large projects. **5. Spring clamp:** Jaws snap closed when the handle is released. Easy to use and great when you're in a hurry. **6. C-clamp:** Versatile and useful in tight spaces, they range in size from 2 inches to more than a foot. **7. Ratchet band:** Best for looping around several items or oddly shaped projects, like chair backs or columns.

Wood hand-screw: #3-0, 6" jaw length, $17.03; #0, 8" jaw length, $20.36; #1, 10" jaw length, 6" opening capacity, $23.31; #2, 12" jaw length, 8½" opening capacity, $26.73
Adjustable Clamp Co, 417 N. Ashland Ave., Chicago, IL 60622; 312-666-0640.

One-handed bar clamp: Quick-Grip #00512, $15–18
American Tool Co. 8400 Lake View Pkwy., Kenosha, WI 53142; 800-838-7845.

Locking bar: Vise-Grip #18DR, $30–35
American Tool Co.

Quick-action bar: #TG7.016, $40.95
American Clamping Corp., Box 399, Batavia, NY 14021; 800-828-1004.

Spring clamp: Quick-Grip #58200, $2.49
American Tool Co.

C-clamp: #120/6, $9.95
Record Tools Inc., 1920 Clements Rd., Pickering, Ont., Canada, L1W 3V6; 716-842-1180 or 905-428-1077.

Ratchet band: #88127, 2", $39.01
Adjustable Clamp Co.

Paint USA Soot & Dirt Remover, #42R, approx. $4.98
Bloch/New England, Box 296, Worcester, MA 01603; 800-344-2171 or 508-754-3204.

White'N Brite tile and grout cleaner, sizes 8 oz. to 1 gal., $9.95–49.95 postpaid
Safechem Chemical Co. Inc., 315 N. Flagler Dr., Suite 300-P, West Palm Beach, FL 33401; 407-832-8800.

cleaning aids

SOOT REMOVER SPONGE

After a fire at his office, Mike Bloch of Bloch/New England, a 110-year-old firm in Worcester, Massachusetts, saw the professional cleaners using spongy orange blocks of natural tree rubber to get soot off walls and woodwork. "It wasn't on the consumer market, so we took it there," says Bloch. Called (straightforwardly enough) the Soot and Dirt Remover, it retails at hardware stores for $4.98. We were skeptical but tried it on a grubby *This Old House* wall. It works. "It's like a giant eraser," Bloch says.

CLEANING TILE AND GROUT

Wonderful stuff, grout: tough, water-resistant, long-lasting. But it's a magnet for dirt and mildew, and horrid to clean. White'N Brite, a potent liquid cleaner with phosphoric acid (don't treat this stuff casually), can restore a 12-year-old shower stall to glacial whiteness. White'N Brite isn't magic: You must scrub it, hard. We found that toothbrushes work best (wear gloves and geek goggles). It works well on unglazed tile floors too—we scrubbed it with a floor polisher.

8 am

8:10 am

cordless drill/drivers

When it comes to drilling holes or driving screws, you can't beat a cordless tool for convenience. "I don't even use a corded model anymore," says *This Old House* contractor Tom Silva. "Unless I've got a pile of drywall to hang or a whole subfloor to screw off, it's faster to grab a cordless and go."

A drill/driver is essentially a drill with added features (including a multi-setting clutch) to handle the heavier stresses of screwdriving. This makes it more versatile than an ordinary drill. The batteries and battery chargers of both tools, however, are a lot better than they once were. Master carpenter Norm Abram's first cordless drill took about three hours to charge and wasn't much better than a Yankee screwdriver for driving screws. These days a drill/driver is ready to go in as little as 15 minutes, more than adequate to keep most people working steadily. Tom, however, packs two batteries on the job. One stays in the charger until the other dies, and he swaps them back and forth all day to minimize downtime. As befits a busy contractor, he owns several drill/drivers with batteries that range up to 14.4 volts. Unless you're using it professionally, though, a 9.6-volt drill or drill/driver will be able to do most anything around your house.

Most cordless drills have a variable-speed trigger switch; slow speeds are for starting holes, fast speeds finish them. Drill/drivers have the same feature but with two separate ranges: a high-speed range for drilling holes and a high-torque range for driving screws. When you shift into screwdriving range, an adjustable clutch inside the tool allows you to set screws at a consistent depth. Most drill/drivers have an electric brake too. When the trigger is completely released, the chuck stops instantly to keep you from overdriving a screw. Whatever its features, a drill or drill/driver should be comfortable to use—but that, Tom points out, is a personal thing. Heft a tool before you buy it, because they vary considerably in weight and balance. The basic types are shown here. Light-duty drill/drivers go for about $60; contractor-grade models are $125 to $300. Standard cordless drills are somewhat less expensive.

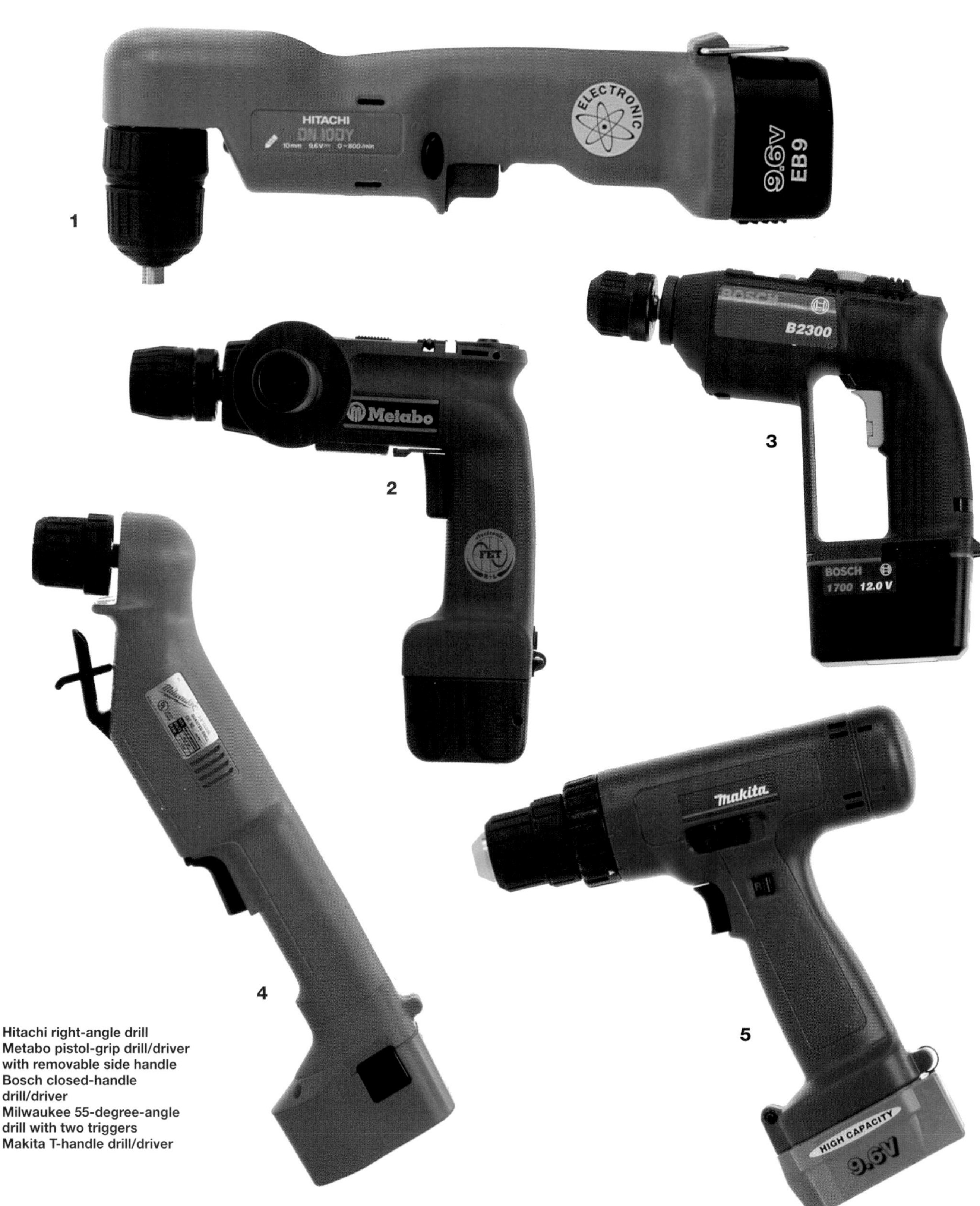

1. Hitachi right-angle drill
2. Metabo pistol-grip drill/driver with removable side handle
3. Bosch closed-handle drill/driver
4. Milwaukee 55-degree-angle drill with two triggers
5. Makita T-handle drill/driver

Makita T-handle cordless drill/driver: #6202DWG, with 9.6v battery, 3/8" keyless chuck, electric brake and adjustable clutch
Makita USA Inc., 14930 Northam St., La Mirada, CA 90638; 800-462-5482.

Metabo pistol-grip cordless drill/driver: #BEAT 100/2, with 9.6v battery, 3/8" keyless chuck, electric brake, two speed ranges, adjustable clutch and additional lightweight screwdriving chuck, with case and battery, #00297, $279; with 1-hr. charger, case and battery, #00296, $339; and with 10-min. charger, case and battery, #00298, $459 (up to 3,000 charges); 1-hr. charger separate, #31.163, $75; 10-min. charger separate, #ICS10, $185
Metabo Corporation, Box 2287, West Chester, PA 19380; 800-638-2264.

Bosch closed-handle cordless drill/driver: #B2300, with 12v battery, 3/8" keyless chuck, electric brake, two speed ranges and electronic adjustable clutch, $286
S-B Power Tool Corp., 4300 West Peterson Ave., Chicago, IL 60646; 800-815-8665.

Milwaukee cordless 55-degree-angle drill with two triggers: #0478-6, with 12v battery, 3/8" keyless chuck, electric brake and one speed range, with 2 battery packs, charger and carrying case, $415 (available in June)
Milwaukee Electric Tool Corp., 131135 West Lisbon Rd., Brookfield, WI 53005; 800-274-9804 or 414-781-3600.

Hitachi cordless right-angle drill: #DN10DY, with 9.6v battery, 3/8" keyless chuck and one speed range
Hitachi Koki USA, Ltd., 3950 Steve Reynolds Blvd., Norcross, GA 30093; 800-598-6657.

ADDITIONAL SOURCES

DeWalt T-handle cordless drill/driver: #DW991K, with 14.4v battery, 3/8" keyless chuck, electric brake, two speed ranges and adjustable clutch, $338
DeWalt Industrial Power Tools, 626 Hanover Pike, Hampstead, MD 21074; 800-433-9258.

CHUCKS

The chuck of a drill or a drill/driver is what holds the bit (or screwdriver tip) in place. It's rated according to the diameter of the largest bit shank it can hold, from 1/4-inch up to 1/2-inch. The 3/8-inch chuck, however, is standard issue on general-duty drills and drill/drivers (Tom wouldn't use a smaller one anyway). The main chuck decision these days is whether or not to go keyless. A simple key with beveled teeth tightens most chucks, and it works fine—if you can find it. "A chuck key is the first thing you lose on a job," says Tom, even if it's clipped to the drill itself. With a keyless chuck there's nothing to lose: A pair of knurled rings around the outside of the chuck lets you tighten or loosen it by hand. "Keyed chucks? I won't miss 'em a bit," Tom says.

CHUCK CHOICES

A chuck with a key (left) is still common, particularly on heavy-duty drills and drill/drivers, but a keyless chuck (center) is more convenient. The keyless variation at right is a quick-connect chuck; it accepts only drill bits and screwdriver tips that have a hex shank.

BATTERIES AND CHARGERS

A battery charger generates heat, and concentrating that heat into the short cycle typical of quick-chargers can damage a battery. The new "smart" chargers get around this by using electronic circuitry to monitor the charging process and minimize heat buildup. Some even refuse to charge a sun-warmed battery until it has cooled. Other than keeping sawdust out, chargers need little care. "I've replaced a few," says Tom, "but that's mostly after something fell on them."

The battery of any cordless tool is actually a collection of batteries, called cells, wired together and fitted into a battery "pack." Each cell is rated at 1.2 volts; join 10 and you have a 12-volt pack. The greater the voltage, the more work the battery can do before it has to be recharged (and the heavier the tool). Battery packs usually last through 800 to 1,000 charging cycles. Replacement packs generally cost $50 to $80. Some battery tips:

- Unlike older nicad batteries, the new nicads can be partially discharged and then recharged without any ill effect. You'll get more life out of one, though, if you recharge it only when the tool starts to feel sluggish. But don't wait until it stops dead—you can damage the cells if you discharge a battery too deeply.
- A battery loses 1 to 3 percent of its charge every day, even if a tool isn't used.
- It's against the law in some states to throw battery packs in the trash; contact the manufacturer for recycling instructions or call the Portable Rechargeable Battery Association at 800-225-7722.
- A few batteries include a built-in LED readout that tells you how much charge is left at any given moment.

TECHNIQUES

Simple things, like drilling a hole perfectly straight, can be surprisingly tricky, just as simple tools are often the ones you have to watch out for. "You might be holding a board for a quick hole and end up drilling through your hand on the other side," cautions Tom, so pay attention to where your hands are at all times.

DRILLING HOLES

STRAIGHT DRILLING

Sight down the drill to line it up with the workpiece, and start the bit slowly. For greater precision, Tom sometimes holds a small square alongside the bit as a guide. "Another way is to have someone else eyeball for you—they can tell you to move the drill one way or another."

step 1.

step 2.

ANGLE DRILLING

Step 1. Drill in straight a half inch or so, then remove and reposition the tool to drill through the side of the starter hole. That gives the bit something to bite into, and it won't wander. If you're using a spade bit, your starter hole has to be a little deeper. **Step 2.** "When you find your angle, stick with it," Tom says. Don't change, or you'll bend the bit and probably ruin the hole. Keep a good grip on the tool to prevent it from twisting. Push the drill firmly into the hole, then pull it out while the bit is still turning.

DRIVING SCREWS

Tom prefers pistol-grip drill/drivers because he can put his muscle directly in line with the bit. He centers his hand high on the tool and pulls the trigger with his lower two fingers. As much as possible, he keeps his forearm aligned with the screw to minimize wrist strain and improve control. To drive a screw with a drill/driver, switch the speed range to low to increase torque, or turning force, and adjust the clutch so it disengages the motor when the screw is fully set (you may have to fiddle with the clutch setting a few times until you get it just right). Then line up the bit with the screw—Tom's 4-inch extension bit gives him something to sight down, making alignment easier to judge. Start the screw slowly, then increase speed and maintain firm pressure to keep the bit from spinning loose. Predrilled screw holes actually make for a stronger connection: The screw shank won't crush wood fibers as it enters, and the threads get a better bite. The pilot hole diameter should be slightly less than the diameter of the screw shank.

hammers

CHOOSING HAMMERS

If you don't have several hammers, you may as well be without tools. Master carpenter Norm Abram has six quite different types in his toolbox. Contractor Tom Silva can get by with three for general contracting, but if he starts working with metal or chipping bricks, he's got to go back to the truck for more.

For every job that requires some sort of pummeling, there's a specific hammer that makes it easier, safer and better done. But even with the right hammer, there's style to consider. When it comes to a standard carpenter's hammer, for example, Norm likes a solid steel model with a leather-wrapped handle. Tom prefers a fiberglass handle with a rubber grip, and *This Old House* director Russ Morash can't stand to swing anything that isn't plain wood. "There is nothing," he says, "as elegant as wood."

SOURCES

Panasonic cordless offset drill/driver: Corner Master #EY6780EQK, with 9.6v battery, ¼" hex quick-connect chuck, electric brake, one speed range and rotating turret, $373
Matsushita Home and Commercial Products Co., 1 Panasonic Way #4A-3, Secaucus, NJ 07094; 800-338-0552 or 201-392-6655.

Extension bits and other screwdriver bits: Various lengths and types can be found at hardware stores and home centers or
Vermont American Tool Co., Box 340, Lincolnton, NC 28093; 800-742-3869.

HAMMER SOURCES

Stanley Tools
600 Myrtle St., New Britain, CT 06053; 800-648-7654.

The Japan Woodworker
1731 Clement Ave., Alameda, CA 94501; 510-521-1810.

Woodcraft Supply
210 Wood County Industrial Park, Box 1686, Parkersburg, WV 26102-1686; 800-535-4482.

Estwing Manufacturing Co.
2647 Eighth St., Rockford, IL 61109-1190; 815-397-9521.

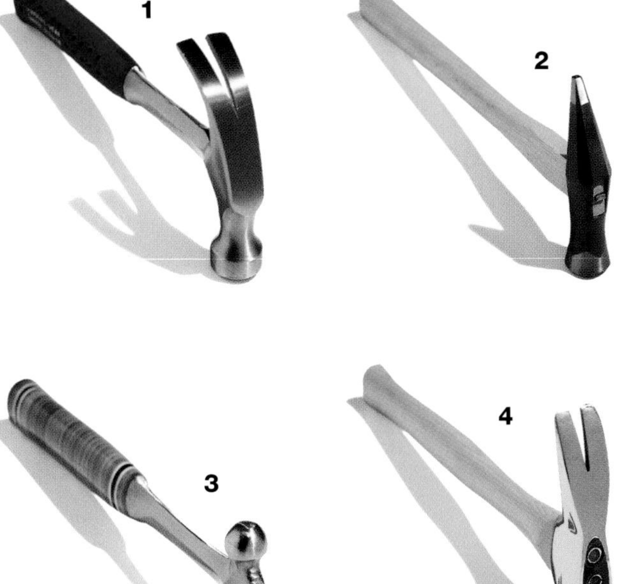

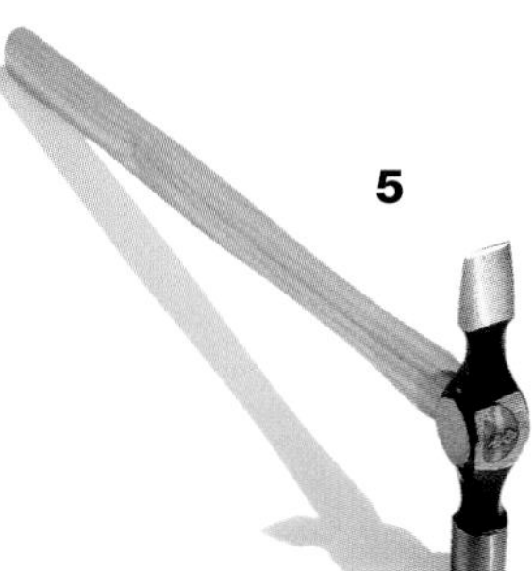

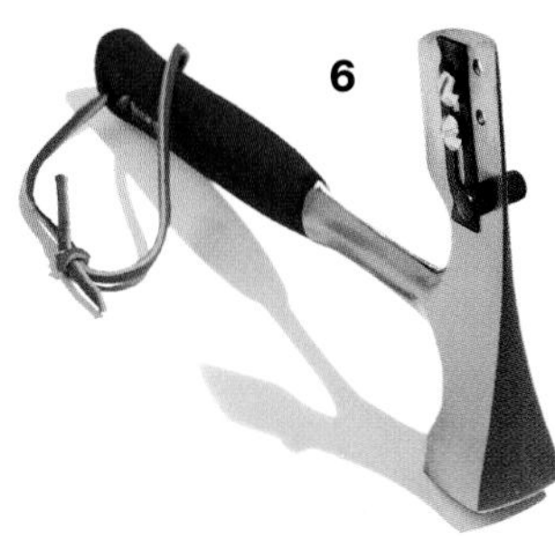

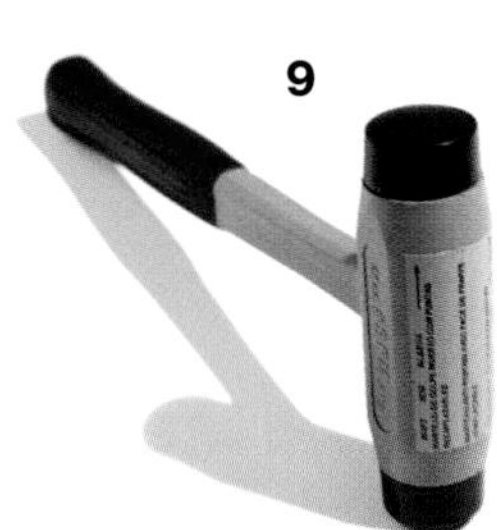

TEN FOR THE TOOLBOX

1. CARPENTER'S

Designed to drive nails, it should never be used to drive metal tools such as chisels or hardened masonry nails. Tom says the best all-around weight is 16 ounces.

2. BOAT BUILDER

The pointed end of the head countersinks fasteners. Imported from Japan, this hammer is a favorite of finish carpenters. The white oak handle is easy on the user's hands.

3. BALL PEEN

Primarily for shaping metal. Can also be used for shaping rivets and driving cold steel chisels and steel punches.

4. RIP

Also known as a framer's hammer. Most have a straight claw that can be used as a last-ditch safety device: A falling carpenter slams it into anything wooden and hangs on tight. Great for tearing up old framing or prying apart two pieces of wood to insert a wedge.

5. TACK

The perfect tool for setting small fasteners. Used by cabinetmakers because the broad face is good for nails and the tapered end is right for striking brads held between fingers. Also good for upholstery tacks.

6. SHINGLER'S

The small, replaceable sliding blade cuts asphalt and fiberglass shingles; the movable pin functions as a gauge. One side of the head is a sharp blade to split shakes, the other serves as a hammer for roofing nails.

7. MALLET

To many, the ultimate driver for wood-carving tools. Usually lathed from a solid piece of wood. Also good for knocking wood-handled chisels and for joinery.

8. BARREL

A fat head puts the hammer's mass near the centerline of the handle, allowing more control when chiseling. Heads are tempered to be soft on the inside and hard on the outside, reducing rebound.

9. DEAD BLOW

Usually filled with lead or steel shot to absorb the impact of the blow and keep the hammer from rebounding. Particularly useful in tight spaces and for assembling furniture. This one has replaceable faces.

10. SPLIT-HEAD RAWHIDE

Faces of water buffalo hide are forgiving. Used to break apart old construction when wood is to be saved or to assemble delicate items. Faces are replaceable. A favorite of timber framers.

TRIANGLE-FACE HAMMERS

Ted Floyd, a carpenter in California, developed the triangular face to better fit into tight spots and to pound close to corners. His finish and framing hammers also have a magnetized T-shaped recess in the cheeks that can hold a nail during the first sideways strike while one hand keeps wood or drywall in place. Their claws have a thinner V, to pull small nails more easily. The hammers come in five sizes, from 10 ounces to 30 ounces. They're very handsome, but there's one big drawback: If your aim is even slightly off, you can leave a sharp dent in your work.

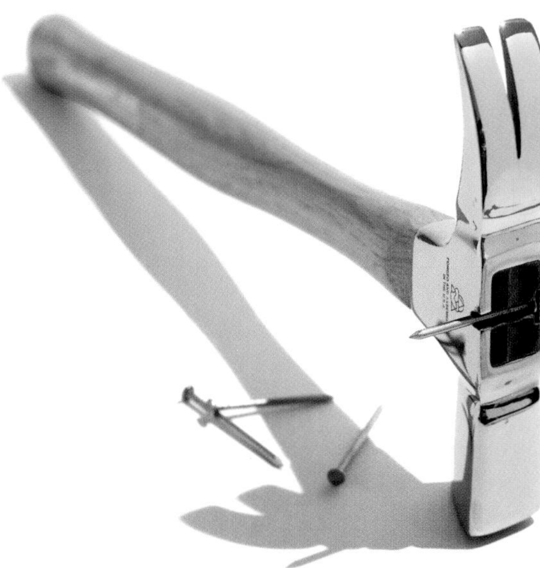

TIP

A tenpenny nail shouldn't require more than two taps to position it, then three solid blows to sink it. A common mistake the neophyte makes is to stand directly over the nail, endangering his head as he raises the hammer. Another frequent error is gripping the hammer too high up the handle. Try this: Stand back, grasp the hammer low and firmly, then swing from the shoulder instead of the elbow, in a full arc. "Above all," Norm says, "remember to keep your eye on the nail."

NORM SAYS

- Look for a hammer with a forged-steel head. Tiny particles of metal tend to chip off cast-steel heads and go flying in all directions.
- Carpenter's hammers with curved claws pull nails better than straight-claw models and help protect amateurs from sticking the claw into their head on the backswing.
- Avoid hammers with waffle-iron faces. A single blow that slips off the nail and onto a thumb is sure to send you to the hospital.
- If a wood, fiberglass or leather-clad hammer handle is slippery, rough it up with sandpaper before using it.

SIDE-STRIKE HAMMER

Refinement is the difference between a tool and an instrument. After World War II, framing carpenter John Hart developed the distinctive West Coast Framing Hammer. Now Greg Gossage, a framer himself, has tweaked Hart's design to make it fit his needs. The scars on Gossage's hammer showed he spent a lot of time pounding sideways—when toenailing, for example. So he redesigned both standard and California-style hammerheads to do the job right. He leveled, checkered and hardened the cheeks, which are usually angled, smooth and soft. The result is the Gossage Side Strike Hammer, which comes in three head weights (17, 22 and 23 ounces) and three handles (polypropylene-jacketed fiberglass, standard hickory and ax-curved hickory).

SOURCES

Triangle-face hammers
Ted Hammers Inc., 6152 Mission Gorge Rd., Suite G., San Diego, CA 92120-3146, 800-645-2434.

Gossage side-strike hammer, $25-30
3771 Porter Creek Rd., Santa Rosa, CA 95404; 800-784-8850 or 707-578-8858.

measuring tools

LEVELS

Somewhere out in the space directly in front of master carpenter Norm Abram hangs a perfect horizontal line. The line is imaginary, but Norm wants to find it anyway, because he's putting up a kitchen cabinet and he doesn't want the dishes to go sliding. "Of course, I'll never find a true level line," he says. "I'm going to be very close, but even my 48-inch level, over 50 feet, will be off half an inch or so."

Or more: Most 48-inch spirit levels can be off by 1½ inches over 100 feet and still indicate a level plane. "Levels are limited by length," Norm says. "If you're leveling nice and straight 12-foot deck joists, even a 28-inch level will do. But if you're setting grade for a 90-foot foundation wall, you need a transit." Norm uses his level mostly to find plumb (vertical) lines. "A living-room floor that goes for 30 feet isn't that uncommon," he says, "but a wall is rarely more than 10 feet high."

WHAT'S INSIDE THAT VIAL?

Spirit levels used to be called "whiskey sticks" because the glass vials were—and still are—filled with alcohol. A dye, usually green, is added to make the bubble inside the vial more visible, especially outdoors. Vials are either injection-molded acrylic plastic, milled from a block of solid acrylic or made of Pyrex glass. The glass vials are often favored for being more consistent—they tend to expand and contract less with variations in temperature. Pyrex and acrylic vials are first filled with methanol, then welded shut with heat or ultrasonic sound, though one manufacturer still caps glass vials with solder, the old-fashioned way.

BUT IS IT REALLY LEVEL?

Accuracy is relative. Many levels use 45-arc-minute vials, which are surprisingly insensitive to minor changes in pitch. (The arc-minute measurement refers to the number of degrees in a circle: 360 degrees in a circle, 60 arc-minutes in each degree, so 45 arc-minutes is ¾ of a degree.) In a 45-arc-minute vial, the bubble won't move unless the level is tilted at least ¾ of a degree. The smaller the arc-minute number, the greater the sensitivity. Look for a level that has at least a 35-arc-minute rating.

Norm likes to use his level to locate studs when installing wallboard. First he makes marks on the floor or ceiling extending from the center of each stud. Then he puts up the wallboard, holds one end of the level at a mark and moves the other end until the bubble is centered in the vial. A line drawn against the level will center each screw in the stud.

LASER LEVELS

There are lots of laser levels out there, but all of them rely on the user to actually get the device leveled, usually with a bubble. Not the tape-measure-size LeveLite, which simultaneously projects two beams—one vertical, the other horizontal—without any adjustment. The secret is its pendulum-hung laser, held plumb by gravity. Just set the level down and a brilliant ruby-red dot appears on the wall at the same height anywhere in the room, a real help when installing cabinets and molding. LeveLite's beams, which are accurate to a quarter-inch over 40 feet, remain level and plumb as long as the case is within 4 degrees of vertical. (If the beam jiggles when the case is tapped, the pendulum is working fine.) Flip the case on its side, and the beams remain exactly 90 degrees apart, useful for tiling floors or laying out foundations. The tool's only apparent drawback is one common to other laser levels: Its beams are hard to see in bright sunlight.

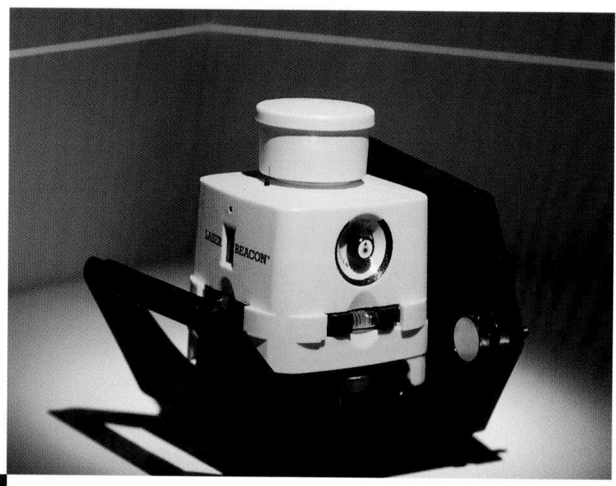

BEACON

The Stanley Laser Beacon creates a continuous line of light vertically or horizontally. It can be hung from a wall bracket to create a level line around the top of a room—for example, to mark the perimeter of a drop ceiling. Or it can be placed in a floor bracket to mark a level baseboard rule or a plumb line on a wall. It can also be mounted on a tripod to create a line around a room for a chair rail. (Don't look directly at the laser, which can cause eye damage.)

LASER/VIAL

A laser built into a spirit level effectively extends its length. Laser levels also tend to have much more accurate bubble vials—down to 5 arc-minutes. The Cuppson On Line Lazer Level combines a laser light with a 32-inch bar. Center the bubble, then push the button to activate the beam. A small red dot appears on walls as far away as 400 feet. It's accurate to within ¼-inch at 60 feet.

VIDEO

Zircon's Laser Vision 6.6 adds sound to laser projection in an 18-inch level. The unit beeps continuously when level, making work easier in confined spaces where bubble vials are difficult to see. A video display tells which direction to move the unit to find level. A push of the "slope" button memorizes any angle so that it can be repeated, useful for establishing pitches on roofs, decks and driveways. The manufacturer claims accuracy to within ⅜-inch at 50 feet.

LeveLite Pocket Lasers: SL indoor unit #20082, two visible beams, $429; SLX2 indoor/outdoor unit #20083, $499; TriLite three-beam indoor/outdoor unit, #20689, $699; SL Interior Pack, #20348, 4 lb., $530; SLX2 Drywall Pack, #20790, 4 lb., $570; SLX2 Contractor Pack, #20347, 4 lb., $600; Tri-Lite Drywall Pack, #0789, 4 lb., $750; Tri-Lite Contractor Pack, #20788, 4 lb., $780
LeveLite Technology Inc., 476 Ellis Street, Mountain View, CA 94043; 415-254-5980 or 800-453-8354.

Stanley Laser Beacon: #42-000, approx. $1,400
Stanley Tools, 600 Myrtle Street, New Britain, CT 06053; 800-262-2161.

Cuppson On Line Lazer Level: #0400, approx. $179
Cuppson Inc., 6506 Headly Court, Levittown, PA 19057; 215-945-0444.

Zircon Laser Vision 6.6: #54033, 18" length, $421.99
Zircon Corp. 1580 Dell Ave., Campbell, CA 95008; 800-245-9265.

Water level by Versa-Level: $49.95
Price Brothers Tool Co., Box 1133, Novato, CA 94948; 800-334-8270.

Cowley Automatic Level: #CL 200, $225
Sonin Inc., Milltown Office Park, Route 22, Suite A202, Brewster, NY 10509; 800-223-7511.

Magnetic Torpedo Level: #991-9, $11.49
Empire Level Mfg. Corp., W229 N1420 Westwood Dr., Waukesha, WI 53186; 800-558-0722.

Glo Lime Line Level: #585, $2.50 for two
Johnson Level & Tool Mfg. Co. Inc., 6333 West Donges Bay Road, Mequon, WI 53092; 414-242-1161.

EYE LEVELS

WATER LEVEL

The age-old knowledge that water seeks its own level applies here. For example, to hang shelves, hang the reservoir on a nail, establish the level of water in the tube at the height you want to repeat, then move the tube anywhere in the room, wait for the water to settle, and mark.

MIRROR TRANSIT

The Cowley level works like a surveyor's transit but is much easier to use. The operator sights a target within the rectangle on the measuring stick and sees a split image like that found in a camera viewfinder. When the target is raised or lowered to level, the split images come together. The unit, which operates on a mirror system, does not have to be leveled to make sightings. It is helpful for laying foundations, checking masonry walls, establishing drainage patterns and leveling large areas such as tennis courts.

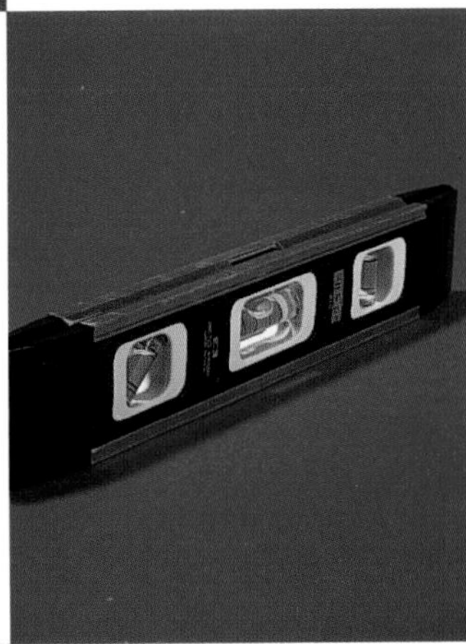

TORPEDO

Torpedo levels are often described as the quickest way to find plumb and horizontal surfaces. Used by plumbers, electricians and homeowners, they are tapered at each end to make them easy to shove into a pocket. Like this one from Empire Level Corp., they are usually 9 inches long and come in magnetic (to stick to pipes or circuit-breaker boxes) and nonmagnetic models. They're perfect for leveling a washing machine or hanging a picture.

STRING

String levels, like this Glo Lime Line Level from Johnson Level & Tool Co., are among the least accurate levels but can be very helpful in special situations. Used by masons to roughly level brick and stone walls, string levels are also useful for homeowners who want to level chalk lines before snapping them. They are useless for finding plumb lines.

NORM SAYS

- Always check a level before buying it. Place it on a flat surface, then level it by adding sheets of paper under one end. Carefully mark where the ends rest, then rotate the level end-for-end. If it still reads level, it's okay.
- Avoid levels with adjustable bubble vials. They seem to get out of whack easier. Buy levels with vials that are firmly glued in or plastered in.
- Treat a level as a delicate instrument. If dropped only once, it may be ruined.
- Always read the bubble straight-on. Reading it at an angle adds to inaccuracy.
- One level is not enough. I use a 48-inch for framing, a 28-inch for checking door and window headers and an 18-incher for tight spaces.
- Levels don't last forever. They do get dropped. Don't try to compensate for a damaged level—get rid of it.

SELF-REELING PLUMB LINES

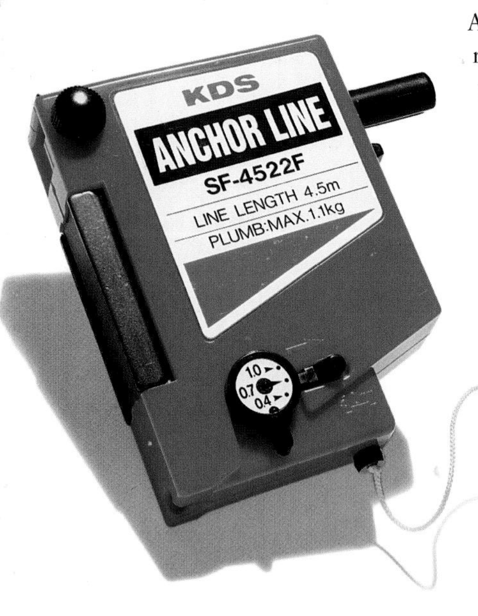

After hanging a door with Plumb-Rite, a self-reeling plumb line, you may decide to abandon your bulky 6-foot level. A spring-loaded pin holds the Plumb-Rite to the side of a wood jamb, then drops an 11-ounce plumb bob exactly 60 millimeters (about 2⅜ inches) from the jamb's edge. Measuring the line's distance from the jamb at any point shows if the jamb is plumb. The Plumb-Rite comes with 14 feet of line. A similar device, pictured here, is the Anchor Line. Both tools have a magnet for plumbing Lally columns or steel doorjambs and a hook for installing cabinets.

TIP

The 12-inch marks on different tape measures may vary by $\frac{1}{16}$ of an inch, mostly because of variations in how the end tabs wobble. This is why it's smart to use one tape throughout a project. And why pros often measure from the 1-inch mark on fine work; just be sure to subtract an inch from the final reading. Buy tapes with hefty, stable tabs.

TAPES AND RULES

At their most basic, measuring devices need be nothing more than sticks of wood on which marks are made. In fact, one way to guard against cutting trim too short or building drawers too big is to avoid using a rule whenever possible. Just hold the piece you need to cut against the space it needs to fit and transfer the measurement directly. But marks on boards don't help builders follow plans or calculate how much wood to buy. Those jobs require the standardized increments of tapes and rules.

NORM ON MEASURING

Ever try to measure inside an opening with a tape measure? Textbooks recommend extending the tape until the tip touches the far side and then adding the length of the case (usually 2 or 3 inches; it's often marked). Norm has found that crooked frames can keep the tape case from fitting into the corner, leading to a short measurement. He just sticks out his tape and lets it curve around one side of the opening. He notes the last place the tape is straight, estimates the width of the gap, then adds the two. For safety he cuts pieces a tad long. "It's easier to trim than to put wood back." Other tricks:

- To mark a line parallel to the edge of a board, lock a fingernail at the desired distance on the rule, lock a pencil against the rule's end and run the whole affair down the board. (You can also use your fingers to lock a pencil a desired distance from an edge. Crease a pad of one finger over the edge and hold the pencil steady as you slide it down the board.)
- On a deck or other framing jobs, ensure uniform stud or joist spacing by tacking a ¾-inch-long scrap to the end of the plate. Hook your tape on the scrap so you can mark each 16 or 24 inches without having to adjust for zero being at the edge, not the centerline, of the first stud. Put an x after each mark and you won't put studs on the wrong side of the line.
- As for shortcuts, ultrasonic and laser measuring tools are fine for estimating, but Norm views a tape with digital display as a crutch that imperils our mastery of fractions.

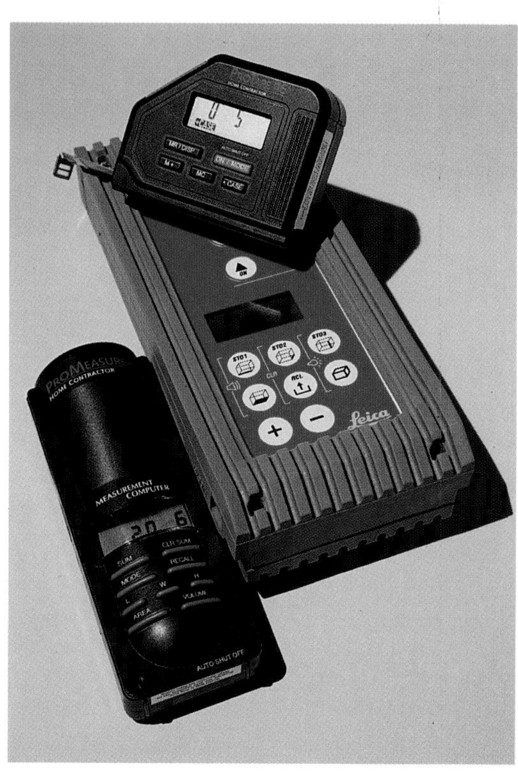

New tools automatically read fractions, convert to decimals and change to metric units. The tape measure at top has digital display. The laser distance meter, middle, and smaller sonic tool, bottom, allow users to stand still while measuring whole rooms.

Tajima Plumb-Rite, #P450Z, $38.95
Nayso Inc, 303 E. Eighth St., New York, NY 10009; 800-229-6770.

KDS Anchor Line, #SF4522, $35.95; KDS 14-oz. plumb bob, $9.75
The Japan Woodworker, 1731 Clement Ave., Alameda, CA 94501; 510-521-1810.

ProTape 16' Home Contractor with digital display: $29.95
Seiko Instruments Inc., 2990 W. Lomita Blvd., Torrance, CA 90505; 800-873-4508.

Disto laser measuring tool: $1,495
Sonin Inc., Milltown Office Park, Suite A202, Rte. 22, Brewster, NY 10509; 914-278-0202.

Ultrasonic ProMeasure Plus Home Contractor: $49.95
Seiko Instruments Inc.

Trigger-lock combination-square head, #PN CS10100 and #PN SC10-036, $150; set with protractor head, combination square and cast connectors, $500; 36" unscaled bar #PN SE10-036, $31.20; 60" unscaled bar #PN SE10-060, $51.50
Super Square Corp., Box 636, Beacon, NY 12508; 800-823-5344.

OVERSIZE SQUARE

In a world of 4x8 plywood and seamless countertops, small marking and measuring tools often fall short. Not the oversize Super Square, which introduces a level of accuracy that would please an aerospace engineer: The advertised tolerance of the heat-treated aluminum I-bars is just .003 inches per foot or less. Cast aluminum attachments slide along or bolt to the I-bars to make a combination square, a protractor or immense right-angle squares. But be warned: These bulky add-ons don't fit easily in tool belts or boxes, and the entire set, complete with a wood carrying case the size of a coffee table, will set you back $500. For between $30 and $50, consider investing in the 36- or 60-inch I-bar alone. These are ideal for cutting plywood, laying tile, checking walls for plumb—any task where an unwarpable, unbendable straightedge is required.

planes

BLOCK PLANES

Norm Abram didn't pay much attention to the block plane that came in the Handy Andy toolbox he was given at age 6. But by the time he was 15, "close enough to an adult to look like I belonged on a job," he had begun to realize how indispensable the tool could be. Working as his father's apprentice during school vacations, Norm pulled out the plane whenever he had to shrink a piece of wood just a bit, whether closing up a gap in a miter joint or squaring up a shingle. Then he discovered the low-angle block plane, a sleeker model with a blade that is angled just 12 degrees up from the sole (instead of the usual 20 degrees) to reduce the chance of ripping deep into the wood. Norm kept his old block plane, "but I'd never use it until the low-angle plane got dull, and I'd push it right to the limit. I think every carpenter does." These days, Norm carries only the low-angle plane in his toolbox. "I like the size of it. It slides easily into my tool belt and fits my hand comfortably so that I can hold a piece of material with one hand and plane it with the other."

Even with his large inventory of power tools, Norm still considers this plane his tool of choice for certain jobs. A tuned-up block plane cuts quickly and accurately. And instead of making clouds of unhealthy sawdust, it produces fragrant, curly shavings. "A lot of carpenters don't use their planes anymore," Norm says. "It's a shame."

Norm likes to use a block plane to knock off sharp edges or to chamfer pieces for furniture, especially when he is working with antique wood. "It looks better than the factory-like finish you get with a router."
This detailing often extends into end grain, which the plane cuts handily. The tool's name comes from its use in smoothing butchers' blocks, originally made with tough end grain.
Norm doesn't bother polishing the sole of a block plane, as some furniture makers do, because one nick from a nail would undo hours of work. But he is careful to polish the back of the iron near the cutting edge, as well as to sharpen and hone the bevel on its front. A blade sharpened only on the bevel side will always be ragged.

Norm polishes the back of the iron by holding it flat against a wet diamond stone and rubbing in a circular motion. To hone the bevel, he uses a store-bought rolling jig to hold the blade steady as he pushes it across the stone, making sure to keep whatever angle is already on the blade. A keen edge can shave hairs off an arm, but a safer test is to try it on a thumbnail. A sharp edge will grab the nail, even if no pressure is applied; a dull edge just skids across.

PARTS AND PIECES

Not only are block planes smaller than most hand planes, but their anatomy is different. The blade on most planes is set at a 45-degree angle and cuts bevel side down. A cap iron on top of the blade has to be flattened just right to prevent shavings from clogging the throat. Block planes are more streamlined. The blade, set at 20 or 12 degrees, cuts bevel side up, like a chisel. Shavings fall away easily without any need for a cap iron.

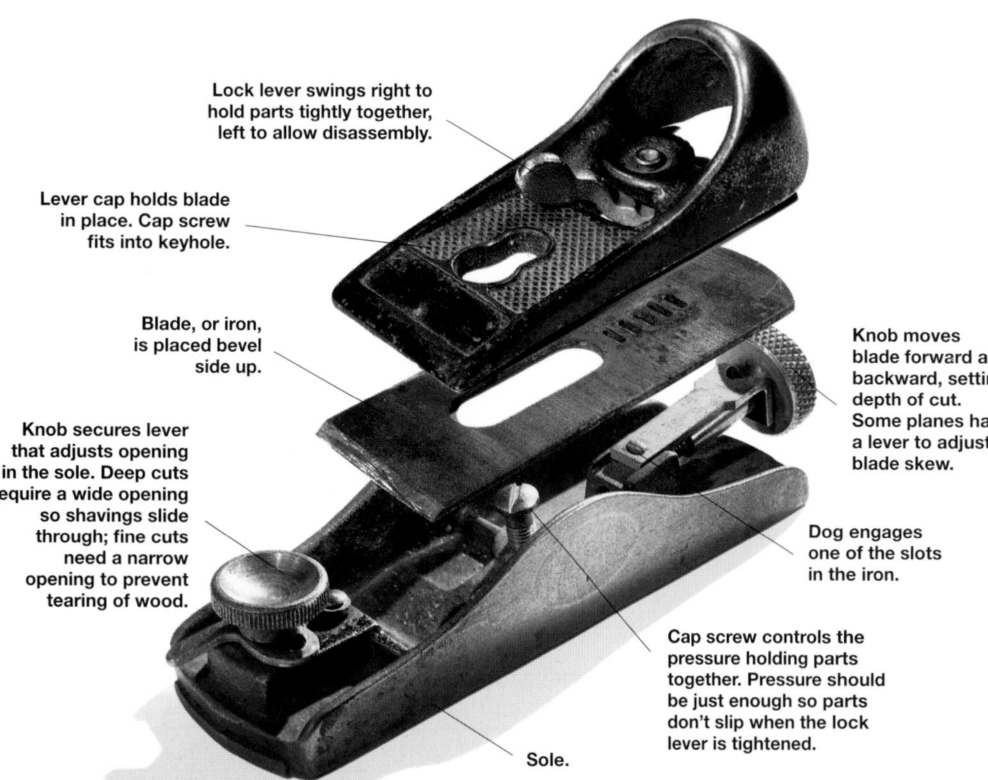

NORM SAYS

- Planes can be tricky, and you can easily ruin a piece of wood by rushing into action without testing first. Use scrap material to test the blade's sharpness, depth of cut and squareness for the kind of trimming or smoothing you want.
- When chamfering end grain, hold the plane at an angle so it cuts down from the board's top face and toward the centerline.
- Don't let resin accumulate on the plane or blade when working on softwood. Rub it off with a rag dipped in a solvent such as turpentine or paint thinner.
- Set the plane down on its side between uses. Laying it upright on its sole can dull the blade.

SOURCES

Block plane: Stanley No. 60½, #17W42, $47.50
Woodcraft, 210 Wood County Industrial Park, Box 1686, Parkersburg, WV 26102-1686; 800-225-1153.

Vise-Type Honing Guide (rolling jig): #60M07.01, $12.95
Lee Valley Tools Ltd., 12 East River Street, Ogdensburg, NY 13669; 800-871-8158.

Diamond whetstone: 6"x2" bench model fine grit, $43.50
DMT Inc., Hayes Memorial Dr., Marlborough, MA 01752; 508-481-5944.

FURTHER READING

Measure Twice, Cut Once: Lessons From a Master Carpenter, by Norm Abram, 1996, 208 pp., $17.95
Little, Brown & Co., 34 Beacon St., Boston, MA 02108; 800-759-0190.

FINESSING JOINTS AND EDGES

With its built-in protection against cutting too deeply, a block plane is perfect for jobs that require fine-tuning. The trick is to take wood grain into account when positioning the tool.

FITTING CORNERS

Cutting a perfect miter isn't easy. The slightest error is doubled when the pieces are mated, and if the surface underneath isn't flat, problems can multiply. For small gaps, Norm avoids using a saw to shave off excess wood. "There's not a tool that can do it better than a block plane," he says. If the gap is on the inside edge, as it usually is, the remedy is to trim a bit from the outside tips. Norm finds it impossible to plane from the tip down; that goes against the grain. He can't plane from the inside up because that would cause the outer edge to tear. Instead, he planes front to back, holding the tool on a diagonal, to preserve a crisp edge.

TRIMMING EDGES

Because a block plane can cut both with the grain and across end grain, Norm uses it to trim a stile-and-rail cabinet door. Planing with the grain of the stile requires little experience, but trimming the tops and bottoms takes skill. At far left, Norm shows what happens if the blade is pushed to the edge of end grain: The wood splits off. To prevent this, he turns the plane around and trims the top rail by pulling the plane toward him. He cuts just the first few inches this way, then flips the plane and finishes the edge by pushing away. When trimming end grain, he angles the plane because the wood shaves more easily. "If you do it at a skew, you're not going right up against that grain." He also finds holding the plane at an angle more comfortable.

SHAVING SHINGLES

Number one on Norm's list of uses for a block plane is trimming wood shingles that are used for siding. Where two shingles meet at a corner, he'll get busy with his block plane to make the edge of one shingle fit the slope of the other. Common irregularities, such as out-of-square shingles, are also easily remedied with the tool. Norm just shakes his head when he sees roofers using electric sanders or grinders on shingles. "They make too much dust," he says.

HAND SCRAPERS

A scraper is the simplest of tools—just a thin piece of steel with a hooked edge. But that hook, pushed across wood, can shave a rough surface smooth in less time than a barber can clear a cheek of two-day-old whiskers with a straight razor.

Scraping shears wood. Sandpaper, by contrast, scours. As sanding proceeds through a series of progressively finer grits, scratches in the wood become more and more shallow until they can no longer be seen. The surface looks smooth, but it may feel fuzzy. A scraped board feels silky, and it shimmers because the wood fibers are sliced, not frayed. The surface undulates like that of a fine antique—a reminder that before the invention of sandpaper in the late 19th century, wood was smoothed with scrapers or hand planes.

Scrapers cannot be used indiscriminately. On pine and other softwoods, scraping may crush fibers or rip them from the surface, leaving pits. But scrapers work on most hardwoods, even highly figured boards such as bird's-eye maple. They make such a shallow cut that there's little danger of tearing fibers, regardless of which direction they face.

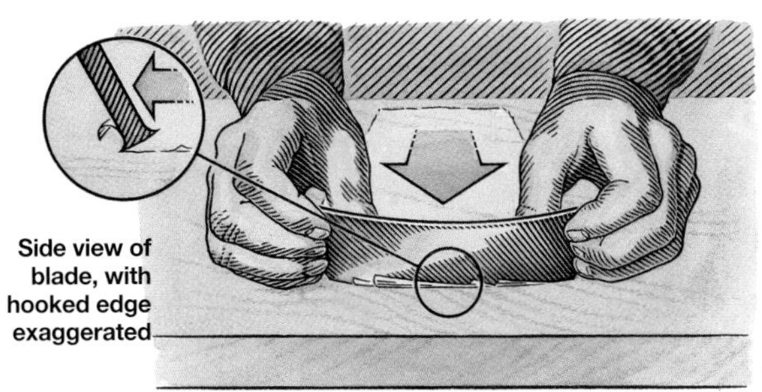

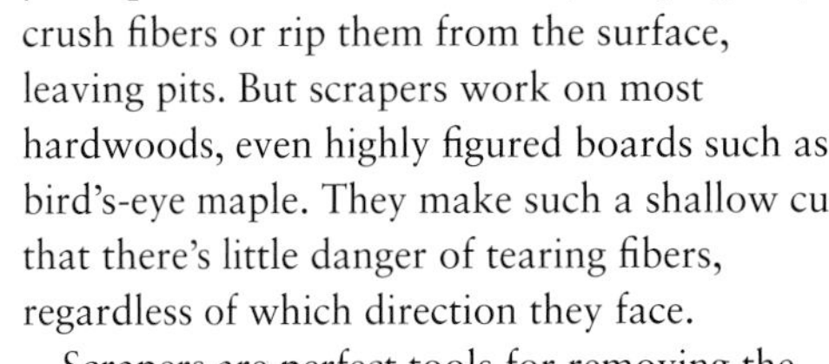

Scrapers are perfect tools for removing the washboard ripples left by the whirling knives of power jointers and planers. If these mill marks lie at right angles to the grain, turning the scraper at a slight skew prevents the steel from following the humps and digging deeper troughs in the wood.

Scrapers can be pulled across a surface, but most woodworkers prefer to push, with their thumbs on the back of the blade, flexing it into a gentle curve. Flexing makes the blade cut at the center, not at the corners, which might gouge.

A scraper should be held nearly perpendicular to the work, tilted just slightly forward. At the correct angle, it will produce lacy shavings. If the tool must be tilted so far forward that the user's knuckles drag on the wood, the hook is too large. A large hook is also inefficient, requiring complete resharpening more often than a small one. When a small hook dulls, it can be restored about a dozen times with a few passes of a burnisher, but a large hook can be reburnished only two or three times.

Using a scraper on bare wood requires neither ear protection nor a respirator. And it even saves money. A package of sandpaper might be gone at the end of a single project. A $12 set of rectangular and curved scrapers lasts a lifetime.

About the only unpleasant aspect of using a scraper is the heat it generates, at times enough to burn thumbs. Scraper holders eliminate the problem but slow down work because the blade must be reset each time it's taken out for sharpening. To protect your skin, slip on leather thumb gloves for long jobs or swaddle your thumbs in masking tape.

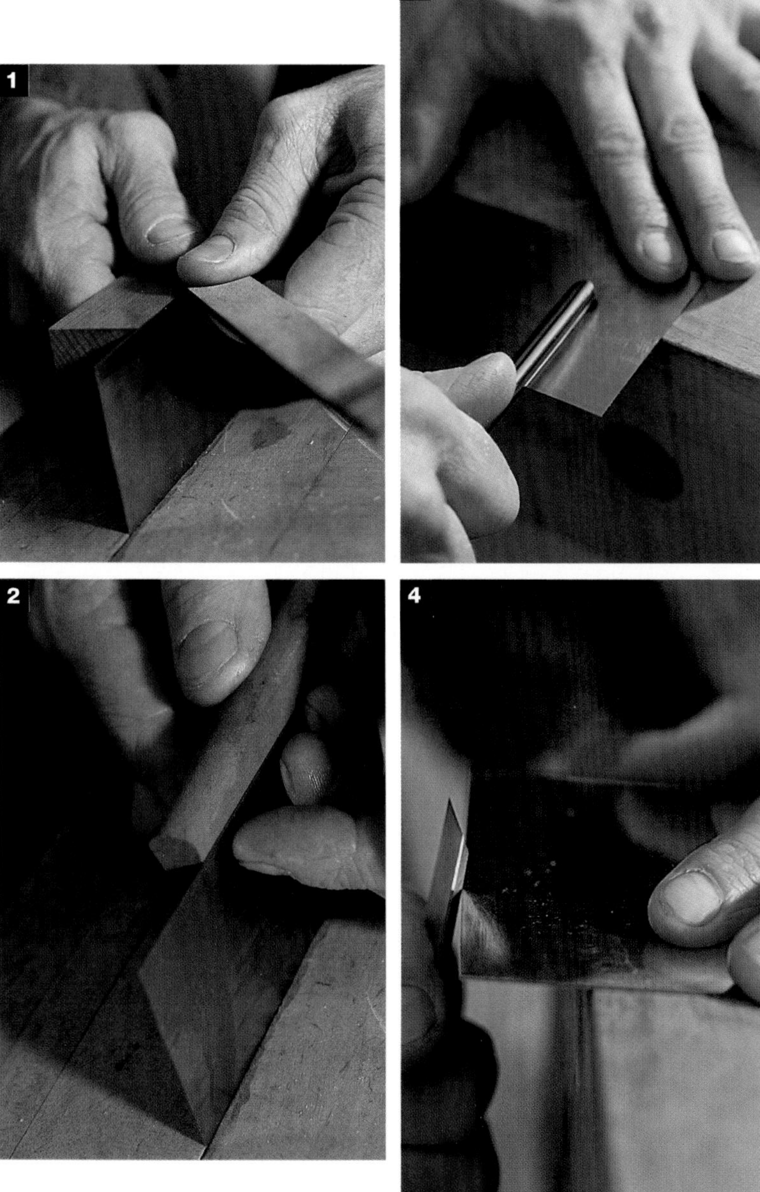

HOOKING AN EDGE

Sharpening a scraper is easy but unorthodox because the aim is not to create a knife edge but to fashion nearly microscopic hooks on both sides of the blade. The process requires a mill file, a sharpening stone, a burnisher and a vise.

1. The first step is to flatten and square the scraper's edges. Woodworker Denis Semprebon secures the scraper in a wood-faced vise and squirts oil onto a single-cut mill file to ensure a smooth cut. He holds a small wooden block under the file and flat against the scraper to keep the file exactly 90 degrees to the edge. Pressing down with both thumbs, he pushes the file along the scraper's edge a few times, repositioning it each time to protect it from uneven wear. "Consistent pressure from beginning to end—that's the trick," he says.

2. The next step is to smooth the sides. Semprebon does this with a small honing stone, which he rubs in tiny circles against one side of the scraper while holding the other side steady so the metal doesn't bend. When the sides feel flat, he polishes the edge with a few passes of the stone, then gives the sides a final touch-up. "The edge has to feel square," he says, testing for burrs with a finger.

3. Fatiguing the steel along the edge makes it more malleable. Semprebon removes the scraper from the vise and rests it against the workbench with one hand as he holds a burnisher flat against the surface with the other. He runs the burnisher across the face of the blade four times.

4. To form the hook, he first smears a drop of oil on the scraper's edge. Then he positions an improvised burnisher—in this case the back of a chisel—perpendicular to the scraper's edge. In one smooth motion, he presses in as he slides along, making a burr. He does this a couple of times, then tilts the burnisher slightly and, with a final pass, curls the burr over into a hook. A tiny hook lasts longest and can be restored many times by quickly repeating steps 3 and 4. The procedure requires finesse, not brawn. "The tendency is to put on too much pressure in all the steps," Semprebon says. A properly sharpened scraper produces gossamer shavings. If it makes sawdust instead, repeat steps 3 and 4. If that doesn't work, the edge probably wasn't square to begin with. The remedy then is to go back to step 1.

Sandvik deluxe scraper: #19K01.04, $10.55; Curved scraper blades for hollow and convex shapes: #70K06.02, $10.50 (set of three)
Garrett Wade, 161 Ave. of the Americas, New York, NY 10013-0459; 800-221-2942.

Veritas triburnisher: #90K03.01, $19.95
Garrett Wade.

10" bastard mill file: #62W13.02, $8.95
Garrett Wade.

Hard Arkansas slipstone: #40M07.01, $15.25
Garrett Wade.

FURTHER READING

The Complete Guide to Sharpening, by Leonard Lee, 1995, 256 pp., $22.95
The Taunton Press, 63 S. Main St., Box 5507, Newtown, CT 06470-5506; 800-888-8286.

THE BURNISHER

A burnisher is little more than a smooth metal rod, but to do its job the rod has to be harder than the metal it's shaping. This burnisher is made to a hardness of Rockwell 62, making it a good match for Semprebon's favorite scraper, which has a hardness of around 50. Unlike some scrapers, Semprebon's comes presquared so the first sharpening is easier.

CUSTOM SCRAPERS

Curved sheets of steel, cut to conform to a specific pattern, can be used to scoop out chair seats, smooth wooden bowls and even remove paint from complex moldings.

GRINDING A MIRROR IMAGE FOR A CUSTOM SCRAPER

1. Semprebon first traces a cutoff piece of scrap molding onto the scraper blank—an old saw blade works fine. If no scrap is handy, use a profile gauge. Semprebon adds slash marks to his tracing so he won't forget which part he wants to remove. **2.** Sparks shoot out as the excess metal is ground away. Spots where the grinding wheel can't reach can be filed. For final passes on fine work, Semprebon adds a hook to the edge, but for rough work, such as removing old paint, he uses the scraper just as it comes off the grinding wheel. **3.** For this job, Semprebon wears a respirator because he suspects the old paint contains lead. He figures that doing the work with a scraper and careful cleanup is safer than using a toxic stripper containing compounds such as methylene chloride, whose vapors can pass undetected through a respirator. Wetting the surface before scraping keeps down dust.

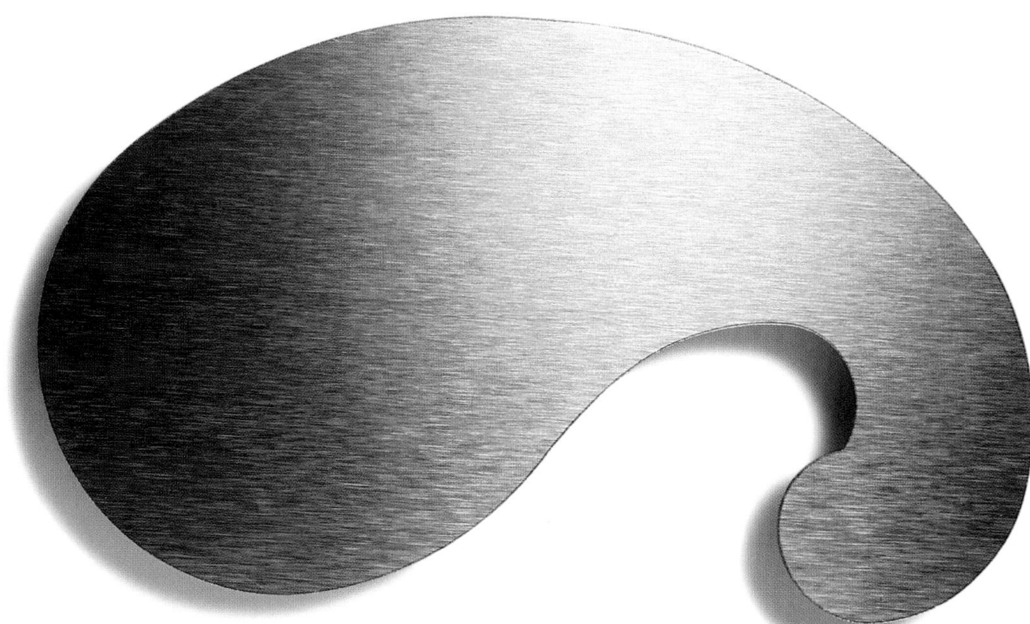

READY-MADE ESSES

A gooseneck scraper is shaped like a draftsman's French curve. Sprung between fingers and thumb, it will conform to a variety of moldings. Leonard Lee, author of *The Complete Guide to Sharpening,* recommends sharpening curved scrapers with a belt sander or a small sanding drum.

TOOL OF A THOUSAND USES

Think of a scraper as a device for removing bits of anything, and there's no limit to the jobs it can do. A scraper can slice off brush strokes before a final coat of paint is applied, bead a board in less time than it takes to set up a router and pare off gummy masking-tape residue.

DECORATIVE EDGES

It's possible to add a bead to a curved edge with a router, but an even crisper detail can be cut with a scraper fashioned from a band-saw blade and fitted into a homemade handle. Semprebon made his own handle from a scrap of rosewood. It has a slot for the blade, a screw to hold the metal snug and a notch to keep the blade a uniform distance from the edge of the workpiece. He first makes a few shallow passes, then loosens the screw and lowers the blade as the cut deepens. A rounded edge does wonders to dress up shelves on a simple bookcase. To patch broken or missing furniture details, Semprebon glues on a slightly oversize piece of wood, then shapes it with a custom scraper. This is far simpler than trying to shape a freestanding patch and then gluing it in place.

routers and carvers

ROUTERS

"It's the one tool I couldn't do without," says Tom Silva of his 1¼-horsepower plunge router. It shapes edges, cuts dovetails, plows grooves and makes molding. When asked about models with soft-start motors and electronically controlled bit speeds, Tom says, "I like my routers plain and simple." But with eight in his stable, the joke at *This Old House* is that he never has to change a router bit—he just changes routers.

Routers come in two varieties: fixed-base and the more versatile plunge type. If you're only going to buy one, here's what to look for in a router:

- 1¼ hp (minimum)
- Both ¼-inch and ½-inch collet capacity
- Plunge capability
- On-Off switch that is easy to reach while holding handles

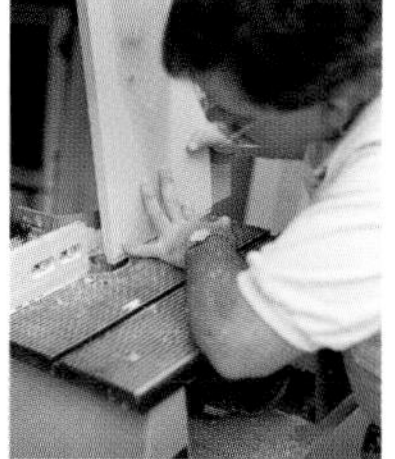

ROUTER SAFETY

A router relies on speed, not strength, to do its work, so kickback is less of a danger than with tools such as the circular saw. Still, a sharpened bit spinning at 23,000 rpm is formidable, so keep fingers well away from the action. That goes for router tables too. Always unplug a router before changing the bit. Because routers kick up a lot of sawdust and shavings, wear a decent particle mask. Though most routers have a chip shield, only fools rely on it for eye protection; be sure you have safety glasses. Protect your ears as well. Tom uses compressible foam earplugs, but muff-type hearing protectors also work if they fit correctly.

On-Off switch
Brush access cap
1½-hp motor
MADE IN USA
Handle
Plunge lever
(on back side)
Plunge post
Collet nut
Collet (inside nut)
Bit
Bearing
Chip shield
WARNING DO NOT REMOVE
Stop nuts
2
1
0
Adjustable depth stop rod
Depth stop control knob
Adjustable
depth stops
Turret
Base

ROUTER PADS

We were skeptical of router pads, those foam squares that supposedly hold wood in place for routing and palm sanding. (For belt sanders, clamps are required.) But when a recent project had us routing the edges of many small hardwood blanks, we tried a Vermont American Bench Vise. Now we're believers. With no clamps to interfere, the job went quickly, and not a piece budged. (Multicolored foam carpet pads or the rubberized antiskid mesh for rugs also work.) The open-weave, 24-by-36-inch Bench Vise lets sawdust filter through, and it can be thrown in the washing machine when it gets dirty.

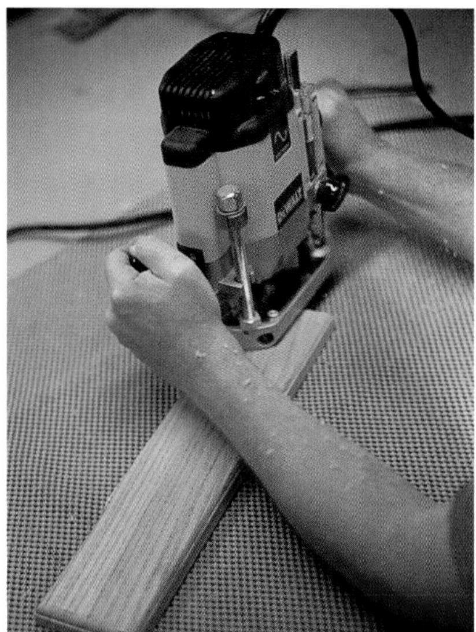

TIP

For a router to cut cleanly, the bit must be sharp and free of nicks. (Carbide bits keep a sharper edge longer than inexpensive high-speed steel bits.) Test a cutting edge by brushing it across your fingernail; if it doesn't grab, the bit's too dull. A quick honing with a diamond whetstone may be all it needs. To protect his bits from damage, Tom sticks them shank first into a block of wood drilled with suitably spaced holes. Clean bits periodically with oven cleaner and a toothbrush and lubricate the bearings; Tom recommends sewing-machine oil.

BASIC GROOVE-FORMING BITS

V-GROOVE

Sometimes called a veining bit, this bit usually has two flutes (cutting edges), available in 60-degree, 90-degree and 120-degree angles. It's used to cut lettering or for general decoration. Changing cutting depth widens or narrows the groove.

DOVETAIL

Though it's often used with a special template to make drawers, Tom uses his—"a lot"—for other joinery. The bit shouldn't be withdrawn from a cut until it exits the edge of the board. Various diameters and angles are available.

MORTISE

Best for routing the recesses (mortises) for hinges and hardware, this bit makes perfectly flat cuts. Cutter diameters range up to 1¼ inches. It's a poor choice for plunge cuts; use a straight bit instead.

ROUND-NOSE

The round-nose (also called a core box bit) can be used to cut the orderly flutes of a classical column. Less formally, Tom has used his to make the chalk rail for a child's chalkboard. The depth of the bit's cut affects the width of the groove.

Router pad: #23468, 24"x 36", $10.50
Vermont American Tool Co., Box 340, Lincolnton, NC 28093; 704-735-7464.

Plunge router: 1½-hp, #693, $328; Router table: #696, $238; Magic Router edge guide: #5043, $67.50; Quicksand random-orbit finishing sander: #333 with dust pickup, $99
Porter-Cable, 4825 Highway 45 North, Jackson, TN 38302-2468; 800-487-8665.

Router bits: Roundover with bearing, #838-817, 1¼" diameter, 9⁄16" length, $34.30
CMT Tools, 310 Mears Blvd., Oldsmar, FL 34677; 800-531-5559.

Wavy Edge ogee, #169-2805, ⅜" diameter, 2 3⁄16" length, $28.99
Eagle America, Box 1099, Chardon, OH 44024; 800-872-2511.

Cove bit, #30-104, 1¼" diameter, 2 3⁄16" length, $35.60; mortise bit, #16-104, ¾" diameter, 2" length, $19.65
Freud, 218 Feld Ave., High Point, NC 27263; 800-334-4107.

Beading bit, #85496M, 1¼" diameter, 2⅝" length, $31.31; core box bit, #85446M, ½" diameter, ¼" shank, ¼" radius, $24.70; rabbet bit, #85151, ⅜" width, ¼" flute diameter, $9.20
Bosch, S-B Power Tool Co., 4300 W. Peterson Ave., Chicago, IL 60646; 800-815-8665.

BASIC EDGE-FORMING BITS

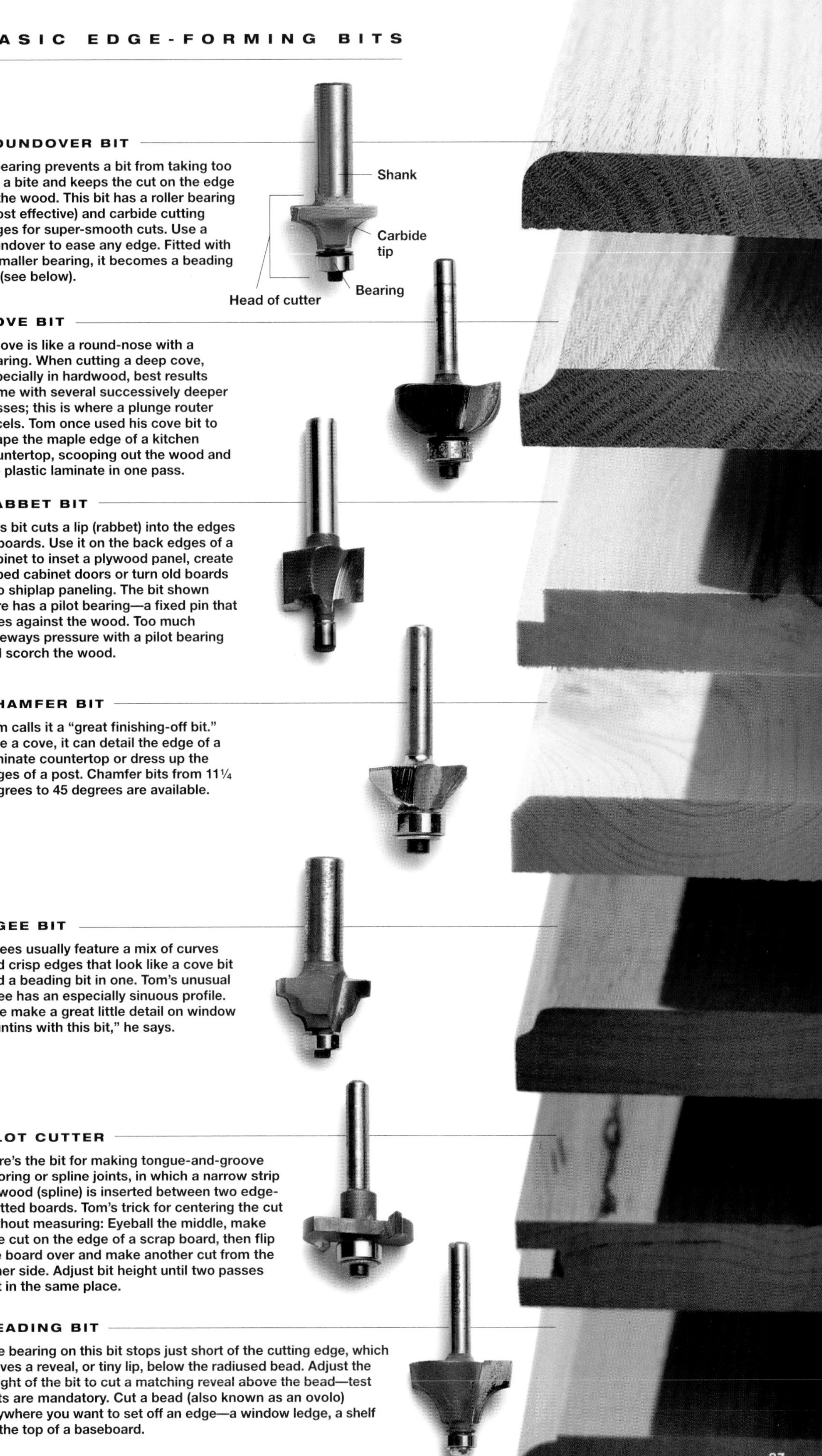

ROUNDOVER BIT

A bearing prevents a bit from taking too big a bite and keeps the cut on the edge of the wood. This bit has a roller bearing (most effective) and carbide cutting edges for super-smooth cuts. Use a roundover to ease any edge. Fitted with a smaller bearing, it becomes a beading bit (see below).

COVE BIT

A cove is like a round-nose with a bearing. When cutting a deep cove, especially in hardwood, best results come with several successively deeper passes; this is where a plunge router excels. Tom once used his cove bit to shape the maple edge of a kitchen countertop, scooping out the wood and the plastic laminate in one pass.

RABBET BIT

This bit cuts a lip (rabbet) into the edges of boards. Use it on the back edges of a cabinet to inset a plywood panel, create lipped cabinet doors or turn old boards into shiplap paneling. The bit shown here has a pilot bearing—a fixed pin that rides against the wood. Too much sideways pressure with a pilot bearing will scorch the wood.

CHAMFER BIT

Tom calls it a "great finishing-off bit." Like a cove, it can detail the edge of a laminate countertop or dress up the edges of a post. Chamfer bits from 11¼ degrees to 45 degrees are available.

OGEE BIT

Ogees usually feature a mix of curves and crisp edges that look like a cove bit and a beading bit in one. Tom's unusual ogee has an especially sinuous profile. "We make a great little detail on window muntins with this bit," he says.

SLOT CUTTER

Here's the bit for making tongue-and-groove flooring or spline joints, in which a narrow strip of wood (spline) is inserted between two edge-slotted boards. Tom's trick for centering the cut without measuring: Eyeball the middle, make one cut on the edge of a scrap board, then flip the board over and make another cut from the other side. Adjust bit height until two passes cut in the same place.

BEADING BIT

The bearing on this bit stops just short of the cutting edge, which leaves a reveal, or tiny lip, below the radiused bead. Adjust the height of the bit to cut a matching reveal above the bead—test cuts are mandatory. Cut a bead (also known as an ovolo) anywhere you want to set off an edge—a window ledge, a shelf or the top of a baseboard.

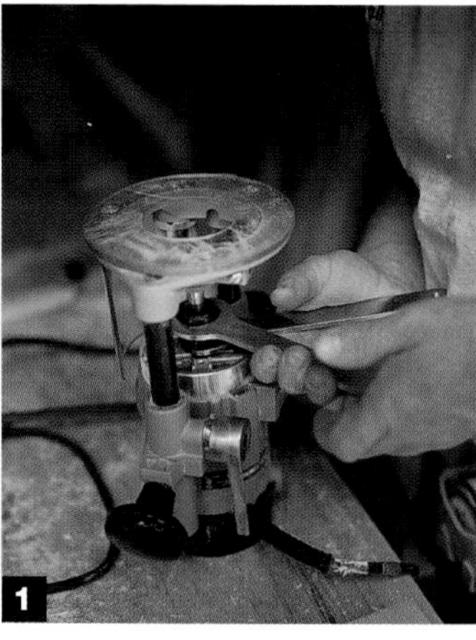

ROUTING AN EDGE

1. Unplug and upend the router. Slip the bit fully into the collet, then back it out 1⁄16-inch or so (to make removing it easier) and tighten the collet nut with the wrenches. **2.** Adjust the height of the cut and lock it in place with the plunge lever. **3.** Move the router counterclockwise around an outside edge so the clockwise-spinning bit bites into the wood. Start and finish cuts halfway down a board; finishing the cut on end grain may cause chipping. At corners, where the router's base has the least support, slow down and keep one handle over the workpiece for good control. **4.** Tom holds his router firmly with both hands and has a steady, balanced stance. "Feel comfortable," he says, "but be ready for anything."

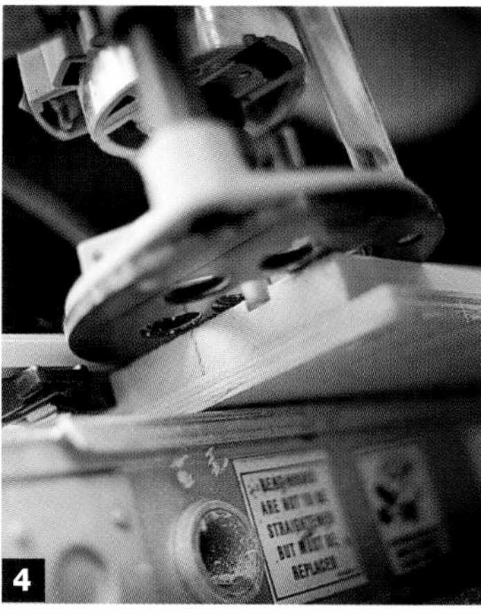

ROUTING A SURFACE

1. "If you don't measure, you can't measure wrong," Tom says. To cut a dado (flat-bottomed groove) for a shelf standard, he first pushes down his unplugged router until the bit just touches the wood. Then, keeping the router in this position, he sandwiches the standard between the depth stop rod and a depth stop on the turret and tightens the depth-stop knob. Now when the router is plunged, the depth of the cut will exactly match the thickness of the standard. **2. and 3.** Always guide the router with a straightedge, or use an adjustable fence attached to the router itself. **4.** When beginning or finishing a cut, move the router slowly and carefully to avoid chipping the wood.

CUTTING A DOVETAIL GROOVE

Here's a trick Tom uses to secure a fixed shelf in a bookcase. **1.** He starts by using a dovetail bit to cut a groove across the width of the cabinet's side piece. **2.** He uses the same dovetail bit mounted on a router table to shape the ends of the shelf. A router table is a useful accessory that holds a router upside down so that wood can be pushed past the rotating bit. **3.** After Tom test-fits the pieces, he assembles the cabinet. A dab of glue at the ends of the groove keeps the shelf from sliding, resulting in an unusually stiff assembly.

Clamp N' Guide fence: #CT50C, 50", $39.95
Griset Industry, 3034 S. Kilson Dr., Santa Ana, CA 92707; 714-662-2888.

Brass bushing kit: #880-002K, $29.90
CMT Tools.

Glue: Titebond II, $23.99 per gal.
Franklin Industries, 2220 Bruck St., Columbus, OH 43207; 800-347-4583.

FURTHER READING

Dictionary of Working Tools, by R.A. Salaman, 1990 (rev. ed.), 546 pp., $27.95
Taunton Press, 63 S. Main St., Newtown, CT 06470; 203-426-8171.

The New Router Handbook, by Patrick Spielman, 1993, 384 pp., $16.95
Sterling Publishing Co., 387 Park Ave. South, New York, NY 10016; 212-532-7160.

Automach HCT-30A, $330
Sugino Corp., 1700 N. Penny Lane, Schaumburg, IL 60173; 708-397-9401.

Ryobi #DC500 with five chisel bits, $65
Ryobi, 5201 Pearman Deary Rd., Anderson, SC 29625; 800-525-2579.

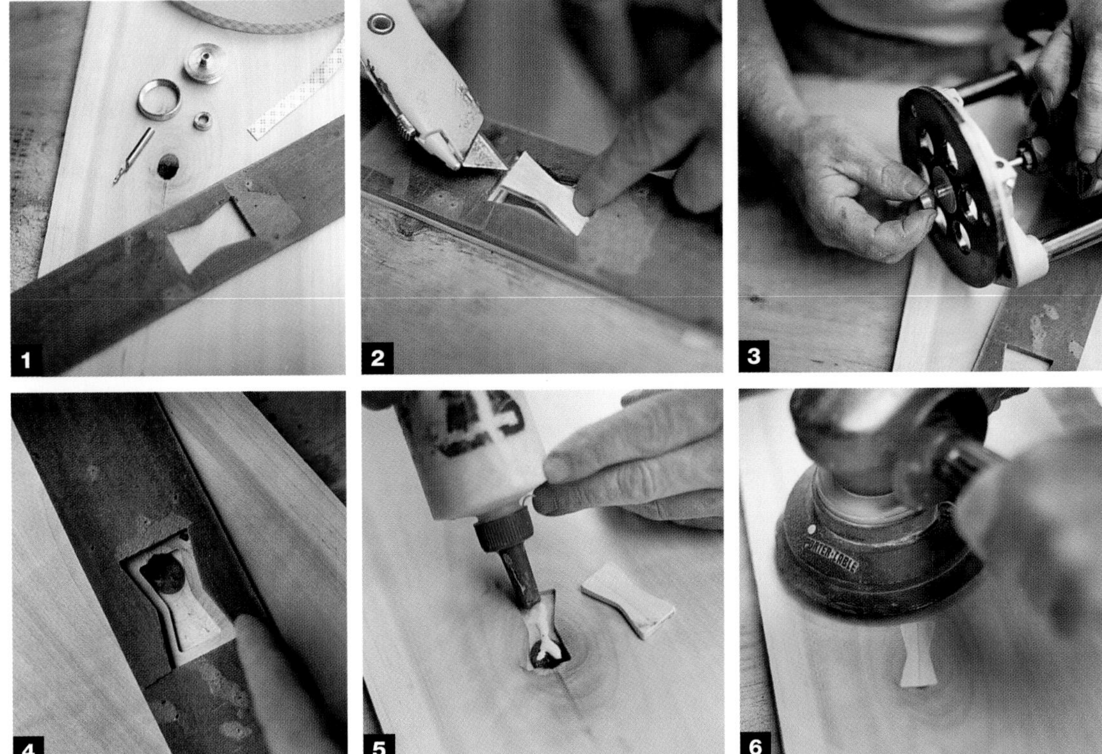

ROUTING AN INLAY OR DUTCH REPAIR

1. With a bushing kit and a hardboard template, you can cut a patch and recess to cover a flaw using the same ⅛-inch bit. **2.** First fit the template guide to the router's base and rout out the bowtie from thin stock. The template (secured with double-sided tape) guides the perimeter of the cut. **3.** Now fit the brass bushing over the guide and set bit depth to match patch thickness. **4.** Secure the template over the flaw. Then rout out a recess for the inlay. **5.** To complete the repair, glue bowtie in place. **6.** Sand it smooth.

DETAIL CARVERS

It's easier than ever to mimic the Old World skills of a woodcarver, thanks to a relatively new kind of tool: the reciprocating carver. Unlike older rotary carvers, which act more like grinders, reciprocating carvers cut just like hand tools, albeit with more noise and vibration. Interchangeable tips match the shapes of traditional carving gouges, chisels and V-parting tools. When we turned on the Automach HCT-30A and the newer, considerably cheaper version from Ryobi, we feared the vibration might make our hands numb after prolonged use. But traditional carving is even harder on the hands than power carving, we concluded after an evening of doing both. Which tool to buy? The $330 Automach runs hotter, but its wider blades remove more wood. More substantial than the $65 Ryobi, it would be the best choice for professionals. But the Ryobi cut well, even in black walnut, and we liked its smaller blades for detail work.

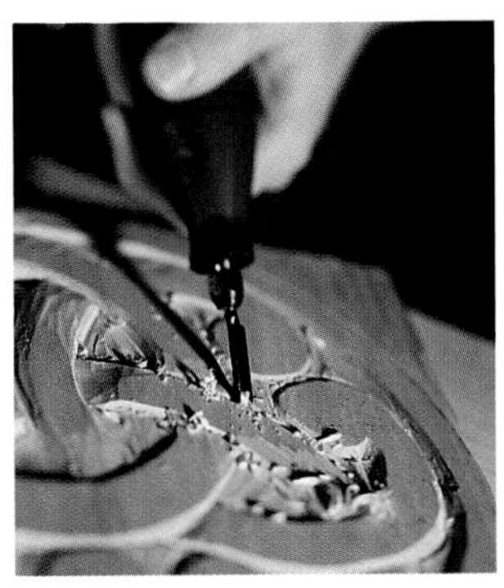

power sanders

DETAIL SANDERS

We edge-sanded more than 60 feet of white oak flooring with the German-made Fein triangle sander and it never once bogged down. Best of all, it doesn't generate much dust. The Fein can be equipped with a variety of accessories—including a scraper, a wood/metal/fiberglass saw blade and a carbide grout saw blade—all interchangeable with a few turns of a hex nut.

There are several cheaper detail sanders on the market. After trying them all, we found only the Bosch B7000 and B7001 came close to the Fein's performance. The Bosch sanders have no fancy accessories, and they vibrate a bit more than the Fein, but they are quieter, lighter and cost about half as much. We liked their click-and-turn pads; no need to hunt for a hex wrench. The B7001 has variable speeds, a big plus for detail work. The Fein is a worthwhile investment if you have big jobs and a tight schedule. For those on a budget with less demanding work, either Bosch tool is a big improvement over fingers and sandpaper.

MINI SANDER

The Nick-Sander uses the power of two AA batteries and the abrasiveness of industrial diamonds to feather the edges of chipped paint, scrub oxidation from electrical contacts and solder joints and even write your name on other tools. (The batteries aren't included, but four diamond tips are.)

SIDING STRIPPERS

To avoid sanding and scraping to get encrusted paint off a house, try the 6-pound Paint Shaver Pro (below). If lead is a concern, the acrylic dust shroud with a vacuum port captures the paint planed off by three spinning carbide "teeth." (But equip your face and vacuum with HEPA filters.)

Porter-Cable's 9½-pound Power Paint Remover (right) is another option. This easy-to-use tool, which whirls a 6-inch tungsten/carbide-studded disk at 3,500 to 4,500 rpm, abrades everything in its path. Use only on lead-free paint; it flings lots of dust.

Neither tool can reach into corners, and both may leave swirls that require sanding. Set any nails before you begin. And as with any machine, stay alert; a moment of inattention can leave nasty gouges.

Fein Triangle sander: model MSx 636-1 (with carrying case), $295
Fein Power Tools, 3019 W. Carson St., Pittsburgh, PA 15204; 800-441-9878.

Bosch 7000, $122; and Bosch 7001, $156
S-B Power Tool Co., 4300 W. Peterson Ave., Chicago, IL 60646; 312-286-7330.

Power NickSander: $9.99
NicSand, Box 29480, Cleveland, OH 44129; 216-351-3333.

Paint Shaver Pro: #PS105TS, $599
American International Tool Industries Inc., 1116-B Park Ave., Cranston, RI 02910; 800-932-5872.

Power Paint Remover: #7403, $284
Porter-Cable, 4825 Highway 45 North, Jackson, TN 38302-2468; 800-487-8665.

BELT SANDERS

Belt sanders are the chain saws of sanding machines: Loud, aggressive and built to remove anything in a hurry. Paused in one place, even for a second, they chew down the hardest wood. *This Old House* contractor Tom Silva warns: "You can do some major damage."

These are simple tools: a trigger switch, a motor, a dust bag and two rollers to guide the revolving sanding belt over a flat base plate. The motor drives the rear roller; the front roller is a lever-tensioned pulley that keeps the belt taut. The motor is usually parked above and perpendicular to the sanding belt (called transverse), but some compact models have in-line motors.

Belt size is the most important feature distinguishing one sander from another. The biggest use belts 4 inches wide and 24 inches in circumference, but there are also 4-by-21, 3-by-24, 3-by-21 and even diminutive 3-by-18 models, as well as a few specialty sanders with belts barely an inch wide. Tom Silva prefers broad 4-by-24 brutes that can rip through decades of paint, grind off nailheads or level wide swaths of wood. These muscle machines are heavy—some tip the scales at up to 15 pounds—but for Tom, that isn't always a concern. He can usually flop work across a couple of sawhorses and let the sander's weight do the work as he steers.

Norm Abram prefers a small sander for its maneuverability. "When you're trying to sand door casings without taking them off the jambs," he says, "a 3-by-21 is nice." It may weigh less than 6 pounds, light enough to sand one-handed. Clamped to a workbench or upended on a special stand, any belt sander can be turned into a stationary tool, good for trimming miters and sizing small pieces of wood.

"The only way to know which one's right for you," Tom says, "is to heft a bunch of 'em." Pick a sander, install the dust bag, then move the tool back and forth as if sanding a tabletop. Tip it sideways and sand the table's edges, then round over a sharp corner. Now try another. Check the feel of the handles as the tool's position changes, and see if the dust bag gets in the way. Install and remove a real sanding belt too; some belt-release levers are uncomfortably stiff.

Belt sanders are first-class generators of sawdust, a known health risk. Dust bags help but can choke trying to keep up with super-coarse belts. Tom hooks his sander to a shop vac whenever possible, even for less demanding applications. He finds vacuum-assisted belts clog less, so they're able to cut faster; they also throw less sawdust into the air.

Before laying a sander on the work, Tom always checks how the belt is tracking over the base plate. If a belt wanders, he slowly turns the tracking knob as the sander is

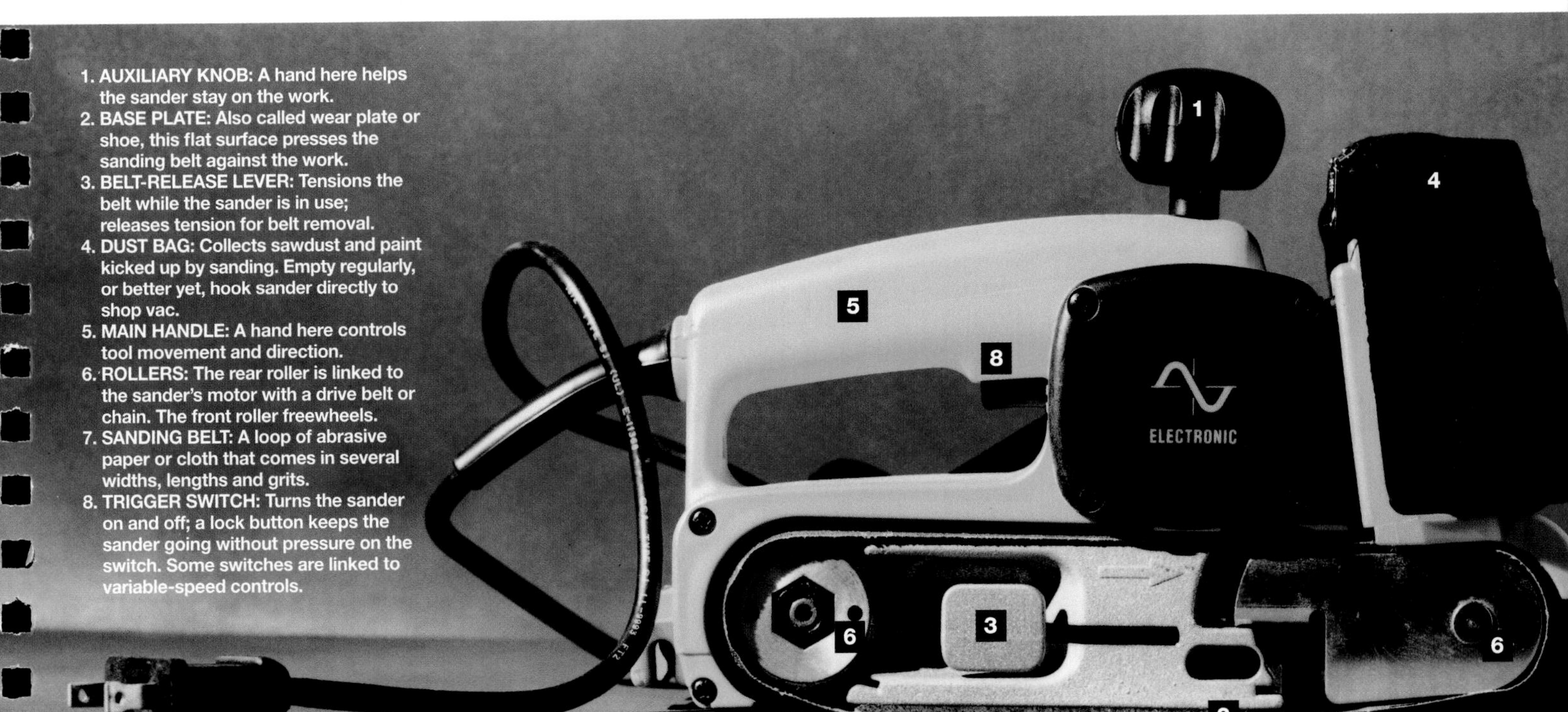

1. **AUXILIARY KNOB:** A hand here helps the sander stay on the work.
2. **BASE PLATE:** Also called wear plate or shoe, this flat surface presses the sanding belt against the work.
3. **BELT-RELEASE LEVER:** Tensions the belt while the sander is in use; releases tension for belt removal.
4. **DUST BAG:** Collects sawdust and paint kicked up by sanding. Empty regularly, or better yet, hook sander directly to shop vac.
5. **MAIN HANDLE:** A hand here controls tool movement and direction.
6. **ROLLERS:** The rear roller is linked to the sander's motor with a drive belt or chain. The front roller freewheels.
7. **SANDING BELT:** A loop of abrasive paper or cloth that comes in several widths, lengths and grits.
8. **TRIGGER SWITCH:** Turns the sander on and off; a lock button keeps the sander going without pressure on the switch. Some switches are linked to variable-speed controls.

running until the belt is centered over the plate and stays there. It isn't necessary to do this often; Tom tracks his tool only when switching belts.

Belt sanders haven't changed much in recent years. Motors are lighter and more powerful, and dust bags have been added. Some sanders now have automatic belt tracking and variable speeds. Tom, who prizes technique over technology, believes belt sanders should be simple and straightforward. "I can't really see the use of more than one speed," he says. His advice: Don't work a surface so hard that it heats up, and don't belt-sand veneer.

Most problems occur when sanders are pressed hard. Norm says, "It takes a fair amount of practice to use a belt sander, because the portion of the belt under the plate is so small in relation to the rest of the tool." But once you know how to drive one, this bulldozer of a tool can be manipulated like a sports car, delivering smooth, stripped wood—fast.

Dust bags on belt sanders make breathing easier. Most mount alongside the tool, but some mount at the top to swivel clear of the work. Many sanders can also be attached to a shop vac.

Intended for close-quarters work and dandy for deck detailing, this sander has a belt only 1⅛ inches wide. A removable handle allows it to sneak into places other sanders rarely go.

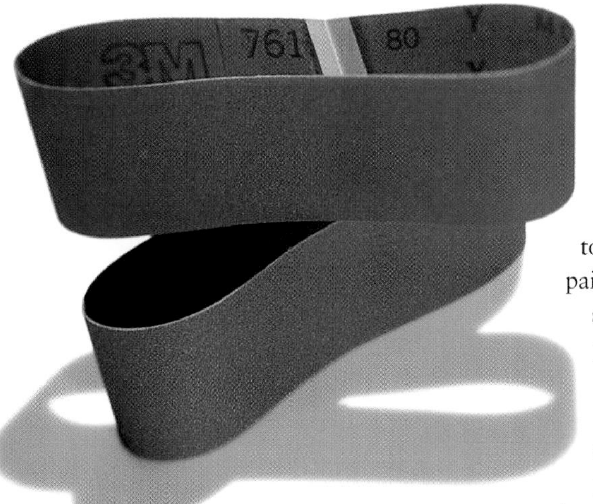

SANDING BELTS

A sanding belt's work is done by legions of tiny, chisel-sharp abrasive granules, or grit, graded by size from super-coarse 24 grit to silky 320. Closed-coat belts pack the grit tightly; they're best for sanding metal and hardwoods. Open-coat belts space out the grit to reduce clogging and so work better on soft, pitchy woods like pine and for stripping paint. Aluminum-oxide grit, typically the least expensive, is good for general-duty wood sanding. New belts with alumina-zirconia (the blue belts) or ceramic aluminum oxide (the purple ones) remain sharp longer but are pricey and hard to find in grits much finer than 120. Tom never uses belts finer than 150 grit; he uses a random-orbit sander for finishing.

All sanding-belt abrasives are embedded in resin atop a backing made of paper or cloth. Paper belts don't last. The best belts are cloth, either tightly woven cotton, polyester or a blend of the two. Polyester is more durable than pure cotton, but belt makers don't always identify the backing.

Until recently, all belts had glued-and-lapped joints that are lumpy. They self-destruct unless they turn in the proper direction (arrows inside the belt show which way they're supposed to rotate). New bidirectional belts have lump-free, taped joints that can be run in either direction. They also last 10 to 15 percent longer and tend to sand smoother than old-style belts. A worn or clogged belt has lousy traction. Tom knows it's time to clean or replace a belt when he doesn't have to rein in his sander as much.

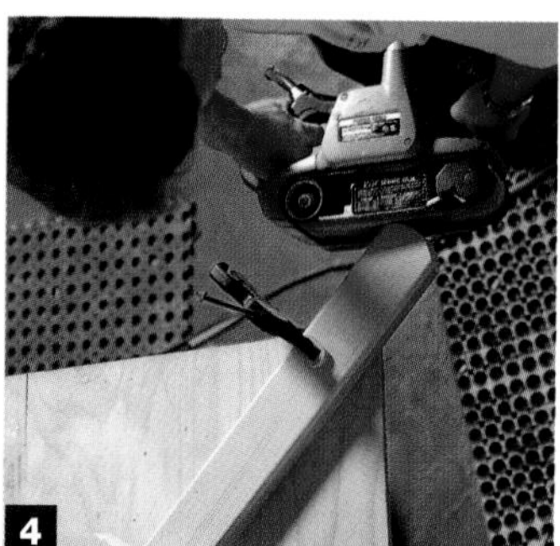

USING A BELT SANDER

1. Stripping paint brings out the best in a belt sander and the worst in belts. Tom Silva keeps his sander moving constantly so as not to overheat the surface and clog the belt. Old paint may harbor lead, so Tom dons a respirator and hooks his 4-by-24 to a shop vac. When he gets down to bare wood, he switches from 36 grit to 120 grit and sands only with the grain. A random-orbit sander with a 220-grit disc finishes the job. **2.** Tom smooths stacks of cedar shelving by pushing back and forth in overlapping passes at an angle to the wood grain. Slight variations in the edge-glued boards can tip a sander and gouge, so Tom skews the tool for greater stability. He'll take out the resulting cross-grain scratches by sanding with the grain. **3.** Smoothing ragged end-grain edges requires an 80-grit belt, solid footing and a good view of the action. To keep the tool from tipping, Tom sands near the belt's centerline, and he watches for a slight shadow between belt and wood, a telltale clue the sander isn't being held flat. **4.** Rounding the ends of table legs calls for light passes and constant motion; otherwise the sander leaves flat spots. Securing the work is a must. A sander can fire loose pieces of wood like a missile launcher.

BELT SANDERS

#DW431, 3"x21", $338
Dewalt Industrial Power Tools, 626 Hanover Pike, Hampstead, MD 21074; 800-433-9258.

#9031, 1⅛"x21", $219-239
Makita USA Inc., 14930 Northam St., La Mirada, CA 90638; 800-462-5482.

#SB10T, 4"x24", $423
Hitachi Koki USA, Ltd., 3950 Steve Reynolds Blvd., Norcross, GA 30093; 800-706-7337.

#362, 4"x24", $245
Porter-Cable, 4825 Hwy. 45 N., Jackson, TN 38305; 800-321-9443.

Belt: 80 grit bidirectional, #761, 3"x21", $2.70
3M, Construction & Home Improvement Markets, 3M Center, 515-3N02, St. Paul, MN 55144; 800-854-4266.

RANDOM-ORBIT SANDERS

Palm-grip: Model #333, 5", single-speed, with dust collection and hook-and-loop pad, $144 (Model #334 identical, but with PSA pad)
Porter-Cable, Box 2468, Hwy. 45 North, Jackson, TN 38302; 901-668-8600 or 800-321-9443.

Right-angle: Model #MSf 636-1, 6", single-speed, with hook-and-loop pad, $745
Fein Power Tools, 3019 W. Carson St., Pittsburgh, PA 15204; 412-331-2325 or 800-441-9878.

In-line: Model #TXE150, 6", variable-speed, with hook-and-loop pad, $281
Chicago Pneumatic Tool Co., Electric Tools Division, 2220 Bleecker St., Utica, NY 13501; 800-243-0870.

MAIL-ORDER SOURCE FOR RANDOM-ORBIT SANDERS AND ACCESSORIES

Klingspor
Box 3737, Hickory, NC 28603-3737; 800-228-0000.

Red Hill Corp.
Box 4234, Gettysburg, PA 17325; 800-822-4003.

RANDOM-ORBIT SANDERS

A random-orbit sander incorporates two simultaneous actions: As the pad spins in circles, an offset drive bearing causes it to also move in an elliptical orbit. The motion isn't truly random, but as you work the two motions overlap, reducing scratching across the grain and keeping any swirl marks to a minimum.

Versatility is another random-orbit hallmark. The tool can strip paint like a belt sander but is easier to control. It can finish like an orbital sander but without grain-direction worries. And because it can suck up and remove dust through holes in the pad, a random-orbit sander is great where ventilation is lousy. Says Norm, "This tool is starting to dominate my sander collection."

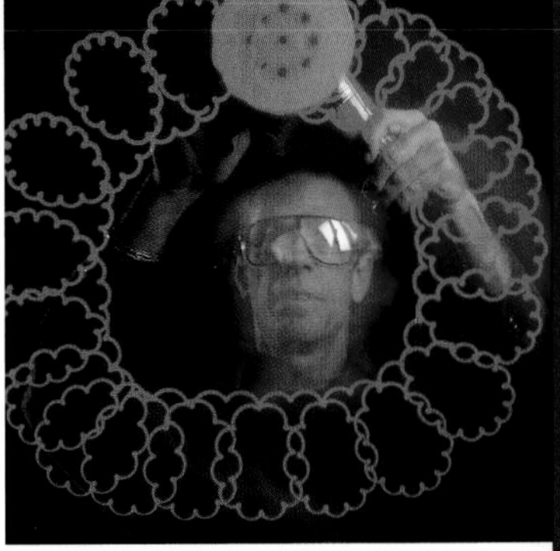

To map a "random" orbit, a tiny battery-powered light was slipped into a felt-covered sanding pad. The photographer snapped this shot as the sander operated against a sheet of Plexiglas.

SANDING TECHNIQUE

When it comes to finish sanding, a random-orbit sander is only as good as your patience. The most common mistake Norm sees is the failure to spend enough time with each grit. You have to start with coarse and earn your way up to fine, but if you don't sand thoroughly at each stage you'll see scratches at the end. Before changing grits, Norm slowly sweeps back and forth across the entire surface with overlapping horizontal strokes, then re-covers the territory with vertical strokes. "I start with 80 grit to strip a finish on solid wood," he says, "otherwise the first step is 100 or 120 followed by 150. I stop there if the surface will be painted. If I'm staining, I go to 180 grit and then 220."

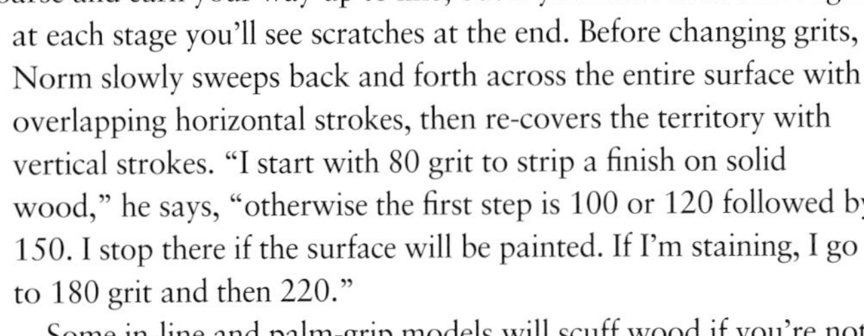

Some in-line and palm-grip models will scuff wood if you're not careful. The pads start spinning as soon as you pull the trigger, but the orbiting motion doesn't kick in until the tool is in full contact with the wood. In that moment of transition, the sander can leave an arc of scratches where it first touches down. Many newer sanders have features that minimize scuffing, but Norm still thinks it's best to start these models while they're flat against the wood.

Because random-orbits generate a lot of dust, most have a bag or canister to collect dust sucked through holes in the sanding pad; a vacuum hookup is even more efficient. Norm's shop vac starts automatically when he turns on the sander.

THE PAPERS

The right sandpaper is crucial if a random-orbit sander is to deliver top performance, but lots of options make choosing tricky. Sanding pads will accommodate hook-and-loop (Velcro-type) or pressure-sensitive adhesive (PSA) discs, but not both. Hook-and loop discs can be taken off and reattached as often as needed; they're the best choice if you change grits frequently. PSA discs are less expensive, but you can't reattach them. Discs come in 5- and 6-inch sizes to match the pad diameter of the sander. Dust collection suffers if the vacuum holes in pad and paper don't line up. Discs can have as few as five holes or as many as 16, so make sure the ones you buy fit your machine. For perfect placement every time (well, almost), Norm sights through the paper's holes as he presses the disc into place. The sandpaper's backing—what the grit is stuck to—is usually paper. Lightweight backings ("A" weight) are best for finish sanding; heavier backings ("C" or "D") are for heavy-duty stock removal or sanding hard surfaces. Unfortunately, manufacturers don't always list backing weight on their packaging.

Uncoated aluminum oxide is the best abrasive for raw wood. On painted or sealed wood, use stearated aluminum oxide discs to minimize clogging.

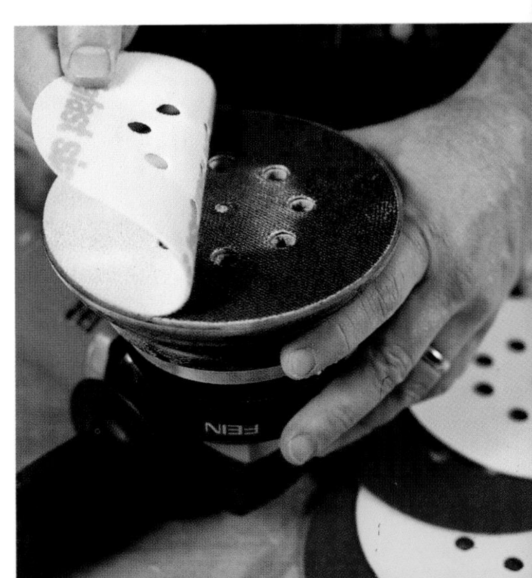

SANDING ACROSS GRAIN IS NO PROBLEM

The eccentric movement of the sanding pad lets you disregard grain direction entirely, a real time-saver on furniture and cabinet projects. When sanding a dresser he built out of recycled antique pine (far left), Norm hardly paused as he crossed the border onto its breadboard edge. His only caution: "Just don't go too far off an edge or you'll round it over when the sander tips." Cabinet doors present a similar situation. The grain changes direction where stile meets rail, but a random-orbit sander zips over the intersection. The wood pegs in this cabinet door (middle) stood slightly above the surface—until Norm took them down with the random-orbit; the tool is unusually efficient at sanding end grain. On narrow surfaces, hold the sander firmly to keep it from sliding off one edge or the other. You'll notice more vibration with a random-orbit sander than with other types. It takes some practice to get the touch just right, particularly with powerful right-angle models. Sanders with variable-speed settings give you the most control. A random-orbit sander won't fit into corners (right), so you will have to finish them off by hand or with a detail sander. Remember to lift, not drag, the vacuum hose over your work. As Norm learned the hard way, the ribs on the hose can rasp away wood edges.

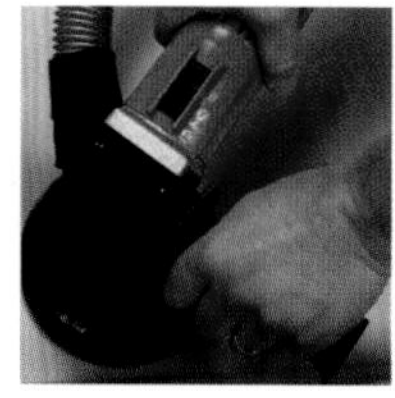

CHOOSING A RANDOM-ORBIT SANDER

Palm-grip models (left) are the lightweights of the family (and usually the least expensive.) They're easy to hold against narrow surfaces, such as cabinet face frames, and they maneuver like sports cars. Pads are 5 inches in diameter; most accept a vacuum attachment. Right-angle sanders (center) have gears that link their powerful motors to the sanding pad. This increases torque and reduces orbiting speed. The tool is a bit noisier (and more expensive) than the others, but you can push it hard without slowing it down—a plus if you're stripping a finish or smoothing a glued-up surface. With a lighter touch and fine sandpaper, it also makes a great finish sander. An in-line sander (right), with its motor directly over the pad, is mechanically identical to a palm-grip sander. In-lines, however, have features that improve their utility, including stronger motors, variable speeds, handles and 5- or 6-inch pads.

handsaws

HACKSAWS

With the right blade, enough muscle and plenty of time, a hacksaw can slice through virtually anything—steel, brass, glass, tile, concrete, ice, bone, even solid stone. Their little piranha-sharp teeth never give up. Plumbers prize hacksaws for slicing through every type of pipe. Butchers use them to make short work of carcasses and frozen meat, and automechanics rely on them to sever rusted bolts. Carpenters sneak hacksaw blades between sash and sills to free windows, and roofers trim gutters with them. Nothing beats a good hacksaw blade for making a smooth cut through hardwood.

Today's frames feature rigid tubular steel with cast-aluminum components, front-end grips, internal storage for spare blades, closed D-handles and 45-degree blade mounts to make angled or flush cuts easier. Some frames adjust to accommodate 10- or 12-inch blades. But the best feature of the newest saws is their lever-action tensioners, which can easily put 30,000 pounds of tension on a blade—the optimal amount for fast, straight cuts, says Paul Gelineau of American Saw & Manufacturing Co. Older stamped-metal frames could theoretically produce 15,000 pounds of blade tension, but only if your fingers were also made of steel and molded to fit the profile of wing nuts.

Despite improvements in frame design, the heart of a hacksaw is its blade. The best are bimetal: a spine of flexible spring steel welded to a toothed strip of hard but brittle high-speed tool steel. Bimetals are tough enough to survive modern high-tension frames. Cheap carbon-steel blades will shatter easily and dull quickly too. Hackers often get into trouble when they put an all-purpose 18-teeth-per-inch blade in a frame and expect it to do everything. That blade works fine on nonferrous metals, metal rods and iron pipe, but on thin-walled tubing such as that used for electrical conduit, the teeth will catch, bind and even break.

Supreme hacksaw with top-mounted tensioning system: #EH-50, $17.50
Estwing Manufacturing Co., 2647 Eighth St., Rockford, IL 61109-1190; 815-397-9558.

Lenox mini-hacksaw: #975, $8.50
American Saw & Manufacturing Co., Tool Div., 301 Chestnut St., E. Longmeadow, MA 01028; 800-628-3030.

6" Pistol-grip hacksaw: #268, $12
Sandvik Saws & Tools Co., 19 Keystone Industrial Park, Box 2036, Scranton, PA 18501-2036; 800-828-9893.

Lenox hacksaw frame: #2012, $5
American Saw & Manufacturing Co., Tool Div.

BLADES

Lenox 10" bimetal: #018-HE, 18 teeth per inch, $1.50
American Saw & Manufacturing Co., Tool Div.

Sandflex 12" bimetal: #3806, 24 tpi, $2
Sandvik Saws & Tools Co.

Lenox 10" bimetal: #032-HE, 32 tpi, $1.50
American Saw & Manufacturing Co., Tool Div.

RemGrit 10" tungsten carbide rod-saw: $3
Greenfield Industries, Disston Div., Deerfield Industrial Park, S. Deerfield, MA 01373; 800-446-8890.

RemGrit 12" tungsten carbide rod-saw: $4
Greenfield Industries.

RemGrit 10" tungsten carbide hacksaw: $4
Greenfield Industries.

RemGrit 12" tungsten carbide hacksaw: $4.50
Greenfield Industries.

A top-mounted tensioning system combines a thumbwheel, threaded rod and an articulated lever arm to make adjustments easy. Change blades carefully: When tension is released, the pins holding the blade often fall out.

Mini or frameless hacksaws work well in tight quarters. Mount blades to cut on the pull stroke to compensate for their lack of stiffness. These frames extend the life of other hacksaws' retired blades, which are seldom worn down on the ends.

Not just a toy for craft projects, a 6-inch pony saw can hack the fine work when a full-size model would be too much tool. This model uses scrollsaw blades with pins.

This saw combines modern features like a chromed-steel tube frame and a textured cast-aluminum handle with yesterday's blade-tensioning: the old-fashioned wing nut.

HACKSAW BLADES

Match the blade to the task, and good results are a cinch. As a rough guide, use coarse blades on thicker or softer materials, fine blades on harder or thinner ones. The coarsest blades, with 14 teeth per inch (not shown), are suitable for aluminum, plastic pipe and wood. **1.** Choose an 18-tpi blade to cut copper, brass and other nonferrous metals, as well as metal rods and cast-iron pipe. **2.** A 24-tpi blade is best for steel conduit and sheet metal no thicker than 5⁄16-inch. **3.** Use 32-tpi blades on thin-walled tubing or sheet metal up to 1⁄8-inch thick or for cutting countertop laminates or plastic. **4. and 5.** Rod and grit saws are the pit bulls of blades. Instead of teeth, they have superhard tungsten-carbide granules that slowly cut through almost anything—brick, stone, concrete or glass block. For straight cuts, use grit saws; rod saws excel at curves.

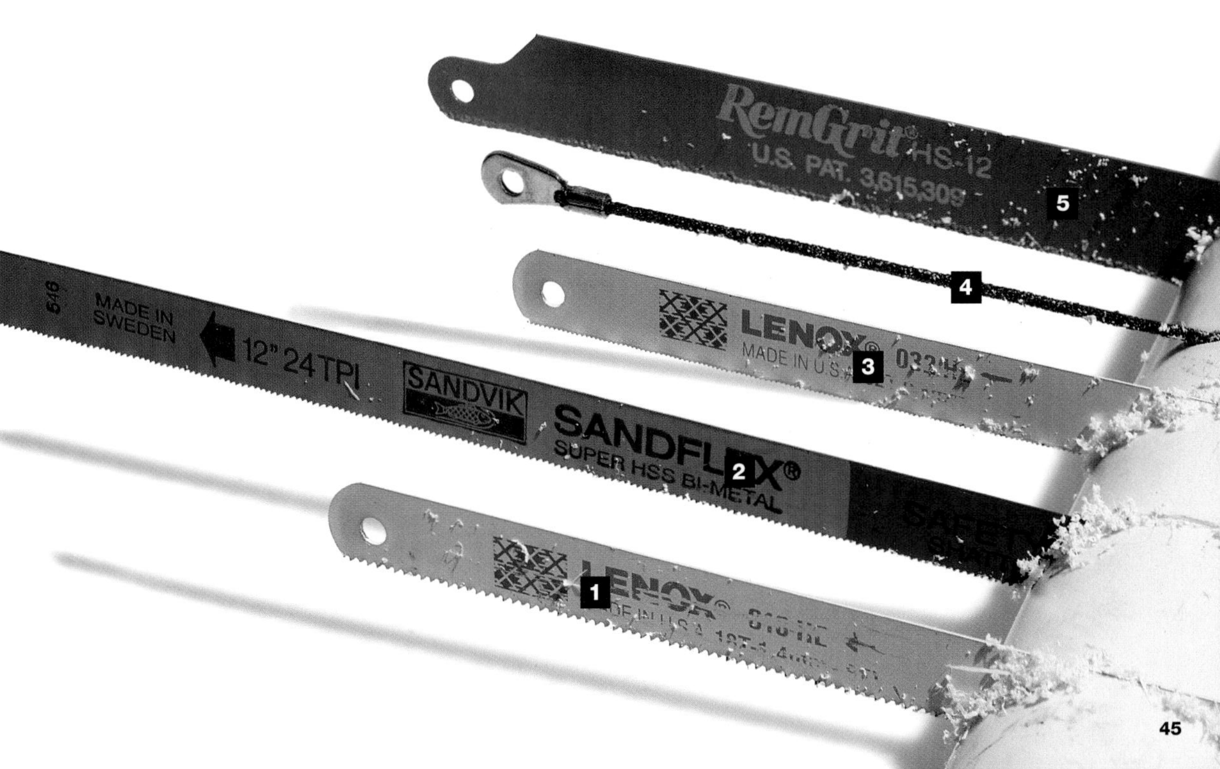

HACKSAW TECHNIQUE

Making the cut: Start by scoring the surface with a file or gently dragging the saw backward a couple of times. Once the kerf is deep enough, apply pressure only on the push stroke to avoid dulling or binding the blade.
Positioning the blade: The teeth should point forward so the saw cuts on the push stroke. Many blades have arrows to show the proper position.
Cutting thin stock and tubing: Use fine-tooth blades and try to angle the saw so at least three teeth are cutting at once. On high-tech alloys, "walk" the blade around instead of pushing straight through. Otherwise, teeth are likely to catch, bind or break.
Sawing slots: Turn a hacksaw into a fast-cutting file by stacking two or three blades in the frame at once.

TIP

Push hacksaw blades too hard, and they'll just call it quits.
Top: Tightening a hot blade can make it snap in two as it cools.
Middle: Overheating a blade can give it a permanent bow.
Bottom: Forcing coarse blades through hard, thin metal may break teeth. Instead of muscling through a cut, ease off and let the tool do the work.

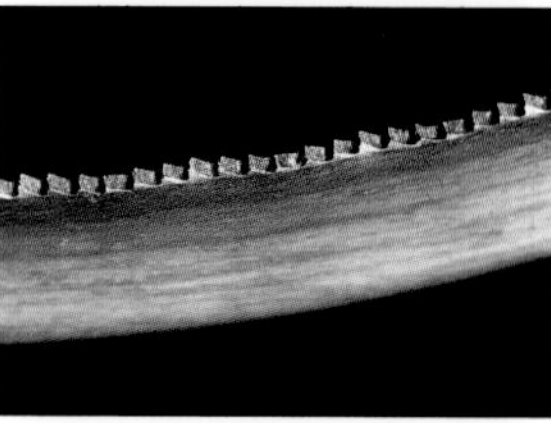

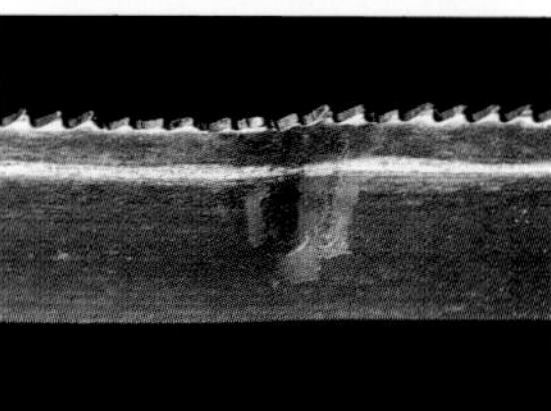

JAPANESE HANDSAWS

Japanese handsaws cut only on the pull stroke, and because a pull stroke puts less stress on the blade, the saws can be made with a harder, though more brittle, steel. This allows for sharper teeth and a thinner blade. A thinner blade makes for an easier cut because less wood is displaced.

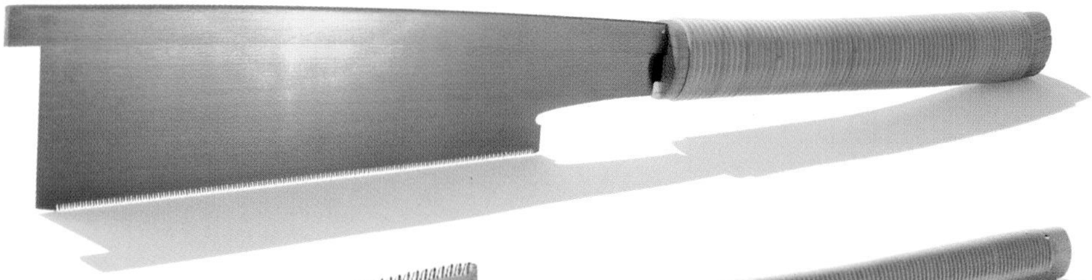

With a blade thickness of only 0.010 inch, the Shindo dozuki is designed for fine cutting of miters and the shoulders of tenons and dovetails. The rattan-wrapped handle enhances grip.

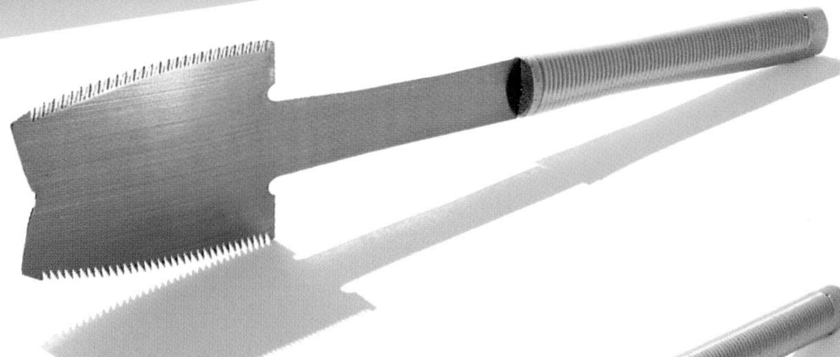

The Azebiki Noko Giri is designed for cutting mortises, making grooves in midpanel and sliding dovetails. It is especially useful when the craftsman must maneuver in a tight area without damaging adjacent work.

This Ryoba Noko Giri has 10 crosscut teeth per inch on one edge, 5 rip teeth per inch on the other. Rip teeth increase in size from the heel (near the handle) to the toe of the blade, while all crosscut teeth are identical.

The Gomuguri is a multipurpose crosscut saw with a less expensive plastic handle and a replaceable blade, which eliminates the need for sharpening.

Shindo Dozuki:
#05.117.18, 8" blade, $225
The Japan Woodworker, 1731 Clement Ave., Alameda, CA 94501; 800-537-7820.

Azebiki Noko Giri:
#15.121.20, 4¾" blade, $24.95
The Japan Woodworker.

Ryoba Noko Giri: #01.111.30, 11½" blade, $85.75
The Japan Woodworker.

Gomuguri: #GNG-265, 10½" blade, $21.95
Tajima Tool Corp., 64 Hill Circle, Waterford, MI 48328; 810-681-6423.

THE QUALITY OF THE FINISH CUT AND EDGES

The blades of Japanese saws generally are thinner and have finer teeth than those on their Western counterparts, so they produce a narrower kerf. When Japanese saws are made, the heat-tempered, high-carbon steel of the blade is scraped so it is slightly thinner at the heel than at the toe and in the center than at the edges. This discourages binding. The difference in teeth is such that lightly running an index finger along the blade of a Western saw should not cause a wound. Doing so along the teeth of a Japanese saw, however, would be much riskier.

The Western saw renders a rough end cut. In the traditional manner, the cut is from one side of the wood straight through to the opposite side.

The Japanese saw leaves a smoother finish. In the traditional manner, the wood is rotated during the cut so each partial cut acts as a guide for beginning the cut on the next side.

TOM SAYS

When using a handsaw, be sure to get your shoulder in line with the saw—and both of them lined up with the cut you're planning to make. That way you can make a straight, clean cut. To those who have nearly forgotten what a handsaw is, get reacquainted. Sometimes it's quicker than a power saw, when you consider the setup time. And it's always a good idea to try your hand at yesterday's skills.

power saws

CIRCULAR SAWS

Norm Abram learned to use a circular saw when he was 14, helping his carpenter father on jobs. "It was the workhorse. It seems we did everything with the circular saw," he says. Norm has since acquired plenty of tools; he uses his saw for its ideal purpose—rough framing. But in a more limited set of tools, the circular saw is still the workhorse.

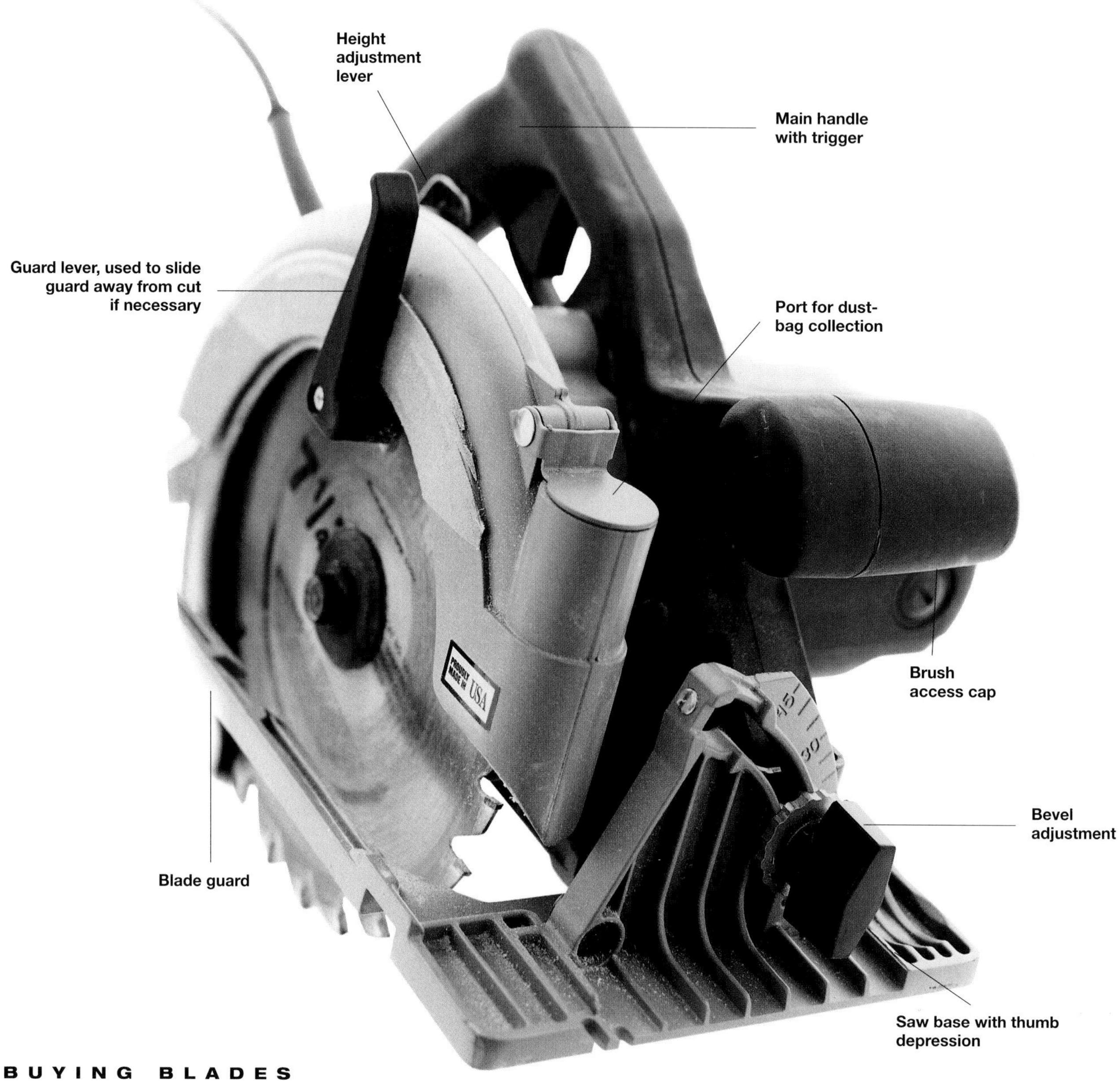

BUYING BLADES

Circular saw blades are usually 7¼ inches in diameter. A decent general-purpose carbide model costs $8 to $10 and will stay sharp much longer than a steel blade. Norm likes thin-kerf blade versions. They cut faster with noticeably less effort. New carbide teeth should be smooth and shiny, not chipped or pitted; check them with a low-power magnifying glass.

CARE OF BLADES

To protect brittle carbide, store each blade in an old mailing envelope or pocket folder (they're great for labeling, too). A dirty blade isn't safe to cut with; use oven cleaner to dissolve gum and pitch after you remove the blade from the saw.

THE BLADES

MASONRY BLADE

A toothless blade, this one is made from an abrasive material that grinds rather than cuts, making it ideal for scoring pavers or for cutting bricks and concrete blocks.

PLYWOOD BLADE

Unlike construction blades that gobble wood, the steel teeth of this blade nibble. It won't splinter thin surface veneers, so it's ideal for cutting plywood paneling and cabinet plywood.

CHISEL-TOOTH STEEL BLADE

This blade probably came with your saw. It cuts fast when sharp and does a decent rip cut and crosscut. Few pros would swap their carbide blades for easily dulled steel blades.

40-TOOTH TRIM BLADE

Premium carbide-tipped blades with 40–60 teeth are finish blades; they cut more slowly—but more smoothly—than similar blades with fewer teeth.

DECKING BLADE

Thin carbide teeth with raised shoulders cut smoothly through pressure-treated lumber and other decking woods. Radial slots in the blade reduce warping.

GENERAL-PURPOSE BLADE

Some blades cut faster, some smoother, but this 20-tooth carbide blade combines speed and long life with a smooth cut. They must be resharpened at saw shops, but a carbide blade is worth it. It's Norm's workhorse.

REMODELING BLADE

Also called a demolition blade. Squared shoulders boost the shock resistance of the few (12 to 14) teeth on this carbide blade; use it on nail-embedded wood.

CROSSCUTTING A 2X4

A few seconds, nothing more—that's what it takes to slice through a 2x4. Do it wrong, though, and you'll remember those seconds forever. Norm's technique may seem overly cautious, but 30 years of accident-free work with the saw is a good endorsement. Proper blade depth is crucial; it minimizes kickback and leaves fewer clawing teeth exposed beneath the cut. **1.** To set the depth, unplug your saw and rest the base on a 2x4 with the blade next to the wood. Pull the guard back so you can see the blade's teeth, then lock the saw at a depth that leaves the lowest tooth no more than about ⅛-inch beneath the wood. **2.** Mark a cutline then saw freehand or (as Norm does) use the square as a guide. **3.** Why guide a simple crosscut? "If you turn the saw even slightly after it's partway through the cut, the blade binds up and the saw can kick back at you." A bonus for beginners: the square also ensures a square cut. **4.** Whatever your method, never lift the saw when the blade is still moving. Instead, push it completely through the wood in a single, fluid motion until the offcut drops free. Generally, Norm says, he keeps the broadest part of the base on the "keep" part of the wood, unless cutting small blocks. (Advice from Norm: if you rock the saw forward and backward slightly as you get ready to cut, you can feel the front of the base slap against the wood; that's when you know the saw is flat and it's safe to cut.)

1

2

3

4

CROSSCUTTING PLYWOOD

Plywood can be more difficult to crosscut than dimensional lumber. The pieces are big, the cutlines are long, and the wood is more likely to splinter. Kickback is a particular risk because plywood flexes readily, so proper support is important. Norm lays several 2x4 supports over his sawhorses; add more if the plywood is less than ½-inch thick. To minimize splintering, cut with the good face down—because a circular saw cuts on the upstroke. **1.** Norm begins by adjusting the blade depth. **2.** Then he sets up a self-clamping straightedge to guide the saw. You can get the same results with a pair of C-clamps and a stiff board. As you cut, be aware that the saw blade will graze the top of each 2x4. **3.** Be sure to keep your free hand well clear of the cut no matter what you're doing. To keep from losing control of the saw, Norm sometimes moves around to continue the cut on the other side of the plywood. **4.** Never lift the saw or pull it backwards while the blade is moving—that's asking for kickback. If the blade binds or your cut goes awry, stop the saw before repositioning it.

1

2

3

4

RIPPING A 2X8

1. To cut a lengthwise strip from dimensional lumber without using a table saw or a rip fence, clamp the board to a sawhorse, then mark the cutline with a combination square. The trick is to move the pencil at the same rate as you move the square. **2.** The line will serve as a rough guide, but the real guide is "the human fence"—Norm's knuckle—which rides along the edge of the wood while his thumb guides the edge of the saw's base. **3.** When Norm reaches a sawhorse, he moves it back slightly before continuing the cut.

1

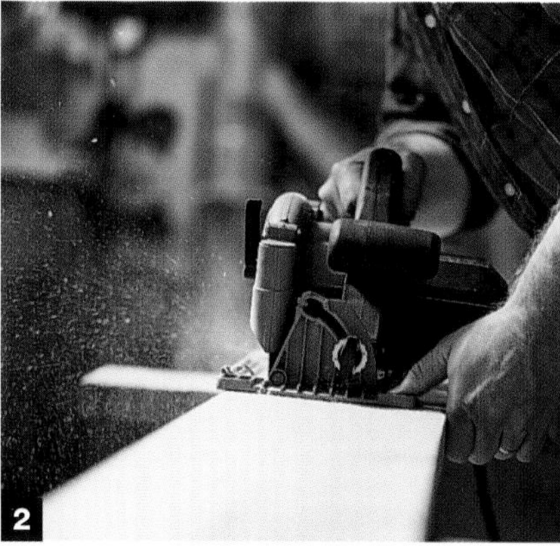
2

3

RIPPING PLYWOOD

When he's cutting a lengthwise strip with a circular saw, Norm supports the plywood on at least four 2x4s laid on top of the sawhorses. They run parallel to the cut, with the innermost pair of 2x4s just inches apart on either side of the cutline. **1.** To prevent the blade guard from "hanging up" on the edge of a thin sheet of plywood, Norm lifts the guard just enough to clear the wood as he starts the cut. **2.** He then releases it as soon as the saw base is past the edge. **3.** To complete the cut, he stops the saw and repositions himself to maintain a safe, solid stance.

1

2

3

$7\frac{1}{4}$" circular saw: Model #447 with electric brake, $245
Porter-Cable, Box 2468 Jackson, TN 38302; 901-668-8600.

Masonry blade: #28001, $6\frac{1}{2}$", $4.55
Vermont American Tool Co., Box 340, Lincolnton, NC 28093; 704-735-7464.

Plywood blade: #25270, $7\frac{1}{4}$", $8.95
Vermont American Tool Co.

Chisel tooth steel blade: #26330, $7\frac{1}{4}$", $4.35
Vermont American Tool Co.

40-tooth trim blade: #15230, $7\frac{1}{4}$", $21.90
American Tool, Inc., 92 Grant St., Wilmington, OH 45177; 513-382-3811.

Decking blade: #26880, $7\frac{1}{4}$", $16.90
Vermont American Tool Co.

General-purpose blade: #27170, $7\frac{1}{4}$", $13.55
Vermont American Tool Co.

Remodeling blade: #27350, $7\frac{1}{4}$", $26.65
Vermont American Tool Co.

SAW SAFETY AND MAINTENANCE

The saw's power and portability are both assets and liabilities. If its blade gets trapped in a cut, the whole saw will kick back violently—at you. Take Norm's advice and always stand to the side of the saw, just in case. Check frequently to make sure small pieces of wood haven't wedged the blade guard open. If the saw doesn't sound right, shut it off and find the problem. It could be a dull or dirty blade. To keep your saw fit, brush off the dust so adjustment levers are clear. If the saw binds or the cuts aren't square, sight down the base periodically to see if it's square and parallel to the blade. Check all knobs and screws regularly to make sure they're tight, and check the electrical cord for fraying. What if you cut through the cord? (Everyone does eventually, Norm says.) Don't repair it; just replace the whole thing. If a saw sputters or sparks, check its brushes. These small blocks of hardened carbon can wear out after a while. Norm's saw makes brush changes easy: He just unscrews each of the access caps, removes the old brushes (carefully—they're spring loaded); and slips in a new set.

NORM SAYS

- Invest in high quality tools. There's nothing more expensive than buying a cheap tool; you'll just have to keep replacing it.
- Test-drive a large tool before you buy it.
- Don't overlook the market in used tools. Good tools have long lives, and many manufacturers can replace broken parts.

RECIPROCATING SAWS

It can cut out, notch out or remove. The reciprocating saw, a brawny tool that has the lineage of a jigsaw and the attitude of a rottweiler, is simple and relatively safe to use, and it's great for reaching the unreachable—wood or metal. Your first cut, though, might surprise you. If it seems like your arms are about to vibrate off, just hold the saw's shoe against the workpiece and push down a bit on the front of the boot; you'll feel the blade bite and the saw simmer down. Tamed.

BASIC USE

Contractor Tom Silva is an old-house guy: He smiles when the walls are open and old wood, sawdust, pipe stubs and piles of stuff are all over the place. Unfortunately, it's hard on his reciprocating saws. "I've gone through a lot of 'em," he says—but when he has to replace one, Tom knows what he wants: variable speed controlled by a trigger. It gives more control over the cut, and, he says, "I really don't want to stop the saw to adjust its speed."

He also likes dual orbital/straight cutting action. On the orbital setting, the blade moves up and down slightly as it goes back and forth, cutting through wood faster. On the straight setting, the blade moves back and forth only; it's better for cutting metal and making fine cuts in wood. Figure on paying $130 to $150 for a reciprocating saw with these features.

Tom favors bimetal blades for most of his work. These blades have flexible spring-steel bodies and hardened tool-steel teeth, characteristics that make them ideal for cutting nail after nail in demolition work. Spring-steel blades cost less but won't cut through more than a nail or two without being damaged.

RECIPROCATING-SAW TECHNIQUE

PLUNGE CUTTING

The key to plunge cutting (in this case, through a plywood subfloor) is to start the saw at a shallow angle using a long, stiff blade. **1.** Rest the saw's shoe against the floor and start the blade slowly; it will scrape, then claw into the wood. **2.** When you feel the blade penetrate, increase its speed and gradually lift up on the handle, pivoting the saw on its shoe. **3.** Maintain a shallow angle until you are sure there are no wires or pipes beneath the floor; then complete the cut with the saw straight up and the shoe flat against the wood.

CUTTING A HOLE

1. After running a couple of screws into the subfloor to steady this small access plate, Tom plunge cuts the hole. **2.** Once the blade gets through, straighten up the saw and twist it slowly but firmly to keep the blade on course. **3.** Follow through with the cut until the scrap drops free. Another way to make this cut is to drill a small pilot hole just inside the guideline before inserting the blade.

CUTTING THROUGH A WALL

1. Cut at a shallow angle to keep wood lath behind a plaster wall from vibrating loose. Any reciprocating saw is a two-handed tool: The trigger hand controls speed and depth while the hand holding the boot guides the blade. **2.** To minimize dust, have a helper hold a vacuum nozzle alongside and slightly below the cutline.

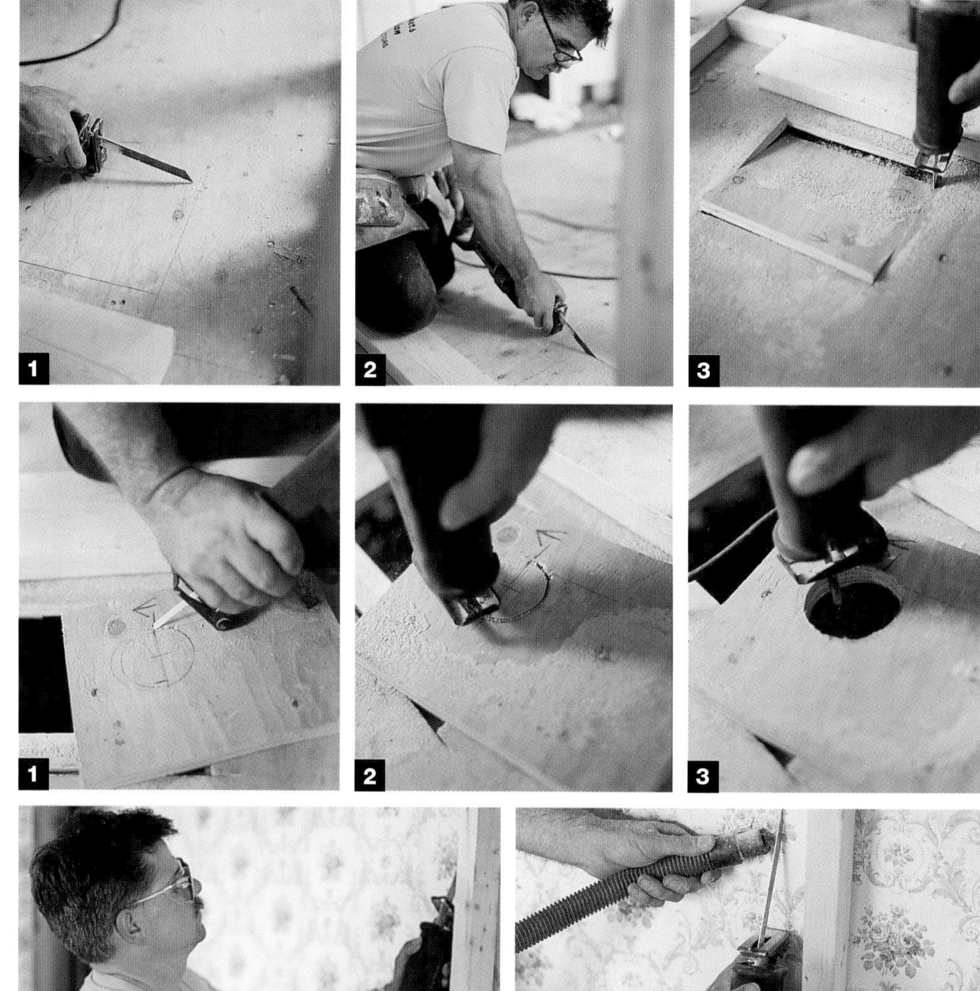

CUTTING STONE

A carbide-grit blade grinds through ceramic and stone (here, a slate shingle). **1.** Tom changes the blades on his saw by turning the spring-loaded locking collar with one hand while slipping the blade into place with the other. Always unplug the saw first. Releasing the collar locks in the blade. (Some saws require an Allen wrench to tighten a clamp around the blade.) **2.** Start with a slow, shallow cut, holding the shoe off the slate to keep from damaging the stone. **3.** Straighten the tool to complete the cut.

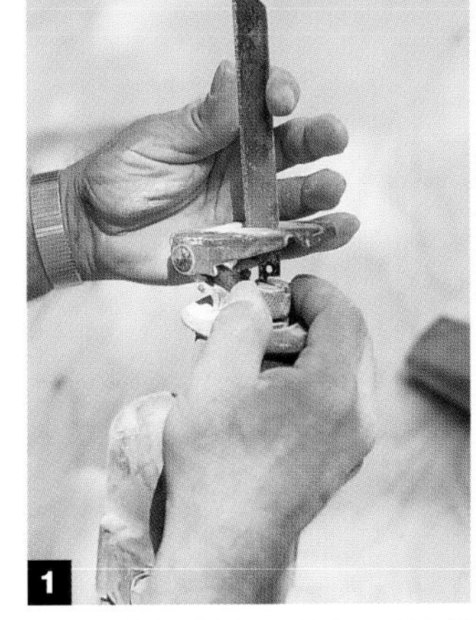

GUIDING A STRAIGHT CUT

1. For very accurate cuts, Tom pins a scrap 2x4 to the floor with his knee or a nail. Holding the saw's shoe and blade against the 2x4, he starts off with a plunge cut. **2.** If he runs out of room, he reverses his grip; here, Tom's right hand pushes while his left operates the trigger and keeps the power cord safe.

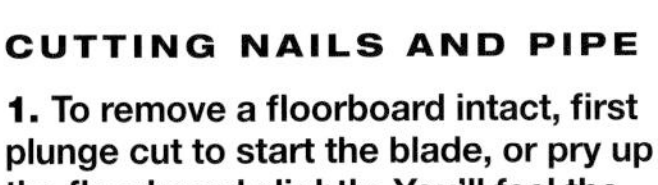

CUTTING NAILS AND PIPE

1. To remove a floorboard intact, first plunge cut to start the blade, or pry up the floorboard slightly. You'll feel the saw slow when it hits a nail, then surge when it cuts through. **2.** Secure pipes so they don't rattle while being cut. Here, an existing notch steadies an old pipe for cutting as Tom pivots the saw on its shoe.

RECIPROCATING-SAW SAFETY

- Always wait for the blade to stop before pulling it from a cut—otherwise the blade will bend and the saw will kick back at you.
- Air rushing out the side of a reciprocating saw blows dust around. Tom always wears safety glasses, and he keeps a dust mask and vacuum ready.
- Slice lumber or pipe so that the kerf (the saw cut) opens up as the blade moves through, otherwise the blade will get pinched and may bend or break.
- A cut pipe is like a sword: Make sure it lands clear of your power cord.
- When cutting wall studs, brace the pieces to keep the kerf open as the cut is completed.

TOM SAYS

- A reciprocating saw can open a wall with ease, but you'd better be careful. Use the angle of the saw to control the depth of the blade.
- Tooth wear is typically greatest within an inch or two of the saw's shoe. To get more life out of a worn blade, adjust the shoe so a slightly different portion of the blade gets more of the action.
- You can turn a long, stiff blade with localized tooth wear into a rather handy handsaw for cutting holes in drywall: Simply wrap the shank with duct tape to make a handle.

THE BLADES

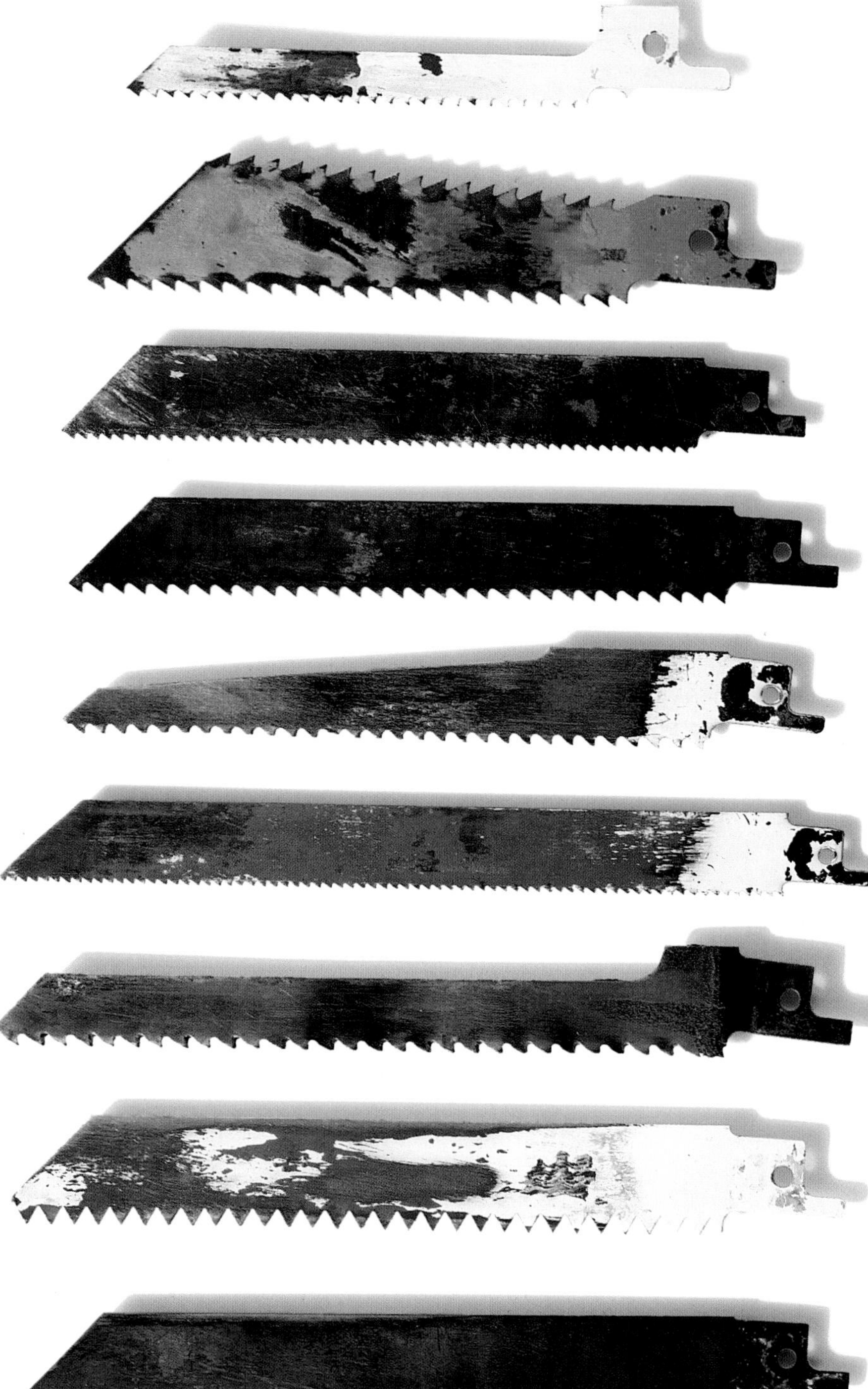

METAL SCROLL BLADE

Cuts curves in metal and wood. (10 teeth per inch)

TWIN EDGE BLADE

A stiff blade for plunge cuts and rough-in work in close quarters. Uneven wear pattern shows the alternating set of the teeth. (7 teeth per inch)

ALL-PURPOSE BLADE

A flexible blade for cutting nail-embedded wood, plastic and nonferrous metals. (10 teeth per inch)

WOOD BLADE

A flexible, general-purpose blade, primarily for cutting wood. (6 teeth per inch)

ROUGH-IN BLADE

A stiff blade for plunge cutting and general use in nail-embedded wood. Note missing and dull teeth; Tom retired this one. (6 teeth per inch)

PIPE-CUTTING BLADE

Cuts plastic and metal pipe, wood, composition materials and nonferrous metals. (10 teeth per inch)

SCROLL BLADE

Cuts contours in softwood and hardwood. Bluish area indicates overheating and loss of temper. (6 teeth per inch)

PLASTER BLADE

Cuts on forward and backward stroke through plaster and through wood or metal lath. (6 teeth per inch)

ABRASIVE BLADE

Cuts rigid fiberglass, ceramic tiles, cast-iron pipe and stone. Edge is lined with tungsten-carbide grit.

WOOD ROUGH-IN BLADE

For general rough-in work in difficult-to-reach places. This blade shows tooth damage but still has some life left. (6 teeth per inch)

THE CUTTING EDGE

It's not unusual for Tom and his crew to blaze through $150 worth of blades in a month; most are either 7-inch all-purpose blades or 12-inch rough-in blades. "I like the finer cut that the all-purpose blade gives on wood, especially plywood," he says, "but the other one cuts faster." He checks used blades for sharpness by pressing a thumb against the teeth. "You'll feel the sharp blades—toss the others." A blade gets hot in use: Wear gloves or let it cool before changing it.

Variable-speed reciprocating saw: #9737, $290
Porter-Cable, Box 2468, Jackson, TN 38302; 800-321-9443 or 901-668-8600 for distributors.

RECIPROCATING SAW BLADES

Metal scroll blade: Milwaukee #48-00-1161, $22.30 for 10
Milwaukee Electric Tool Co., 13135 W. Lisbon Rd., Brookfield, WI 53005; 800-274-9804.

Twin edge blade: Metco #48-00-1131 [discontinued]. A similar blade is Better Tools' "Bore-Hawg" #10202, about $7.99 for 2
Better Tools Inc., 206 River Ridge Circle, Burnsville, MN 55337; 800-798-6657.

All-purpose blade: Milwaukee #48-00-1064, $8.95 for 10
Milwaukee Electric Tool Co.

Wood blade: Milwaukee #48-00-1062, $8.95 for 10
Milwaukee Electric Tool Co.

Rough-in blade: Lenox #S656R, $3.59 each
American Saw & Mfg. Co., 301 Chestnut St., East Longmeadow, MA 01028; 800-628-8810 or 413-525-3961.

Pipe-cutting blade: Lenox #S810R, $3.87 each
American Saw & Mfg. Co.

Scroll blade: Milwaukee #48-00-1041, $13.65 for 5
Milwaukee Electric Tool Co.

Plaster blade: Milwaukee #48-00-1052, $16.50 for 11
Milwaukee Electric Tool Co.

Abrasive blade: Milwaukee #48-01-1420, $17.25 each
Milwaukee Electric Tool Co.

Wood rough-in blade: Lenox #S156R, $5.63 each
American Saw & Mfg. Co.

JIGSAWS

It's a tool of many names—saber saw, bayonet saw—and many uses. The short, stabbing, up-and-down motion of the blade enables a jigsaw to cut big sweeping curves and straight lines or nibble its way into tight corners. *This Old House* contractor Tom Silva grabs one when he needs to scribe trim to brick chimneys, fit cabinets to bowed walls, cut pipe, finish cuts made by circular saws or shape metal thresholds to door jambs. The tool is invaluable for plunge-cutting holes for sinks, pipes and electrical outlets and making decorative scrolls and shelf brackets—jobs where cutting finesse, not speed, is important.

Other tools can do similar work—sort of. Reciprocating saws can cut curves but are too unruly for fine work. Circular saws cut through wood like it's butter, but only in straight lines. A handheld coping saw is precise but slow, and a band saw—though nearly as versatile as a jigsaw—isn't something to carry up a ladder. No wonder the little jigsaw is often the second power-tool purchase a homeowner makes (drills come first).

The biggest decision facing anyone buying a jigsaw is handle style. And on this point jigsaw owners are deeply divided. European carpenters prefer holding their saws by the barrel-shaped housing that surrounds the motor. Americans are equally fervent about the virtues of top-handled models. Yet Tom's a barrel-grip guy. "A top-handle saw feels nice at first, but not once I start cutting. A barrel-grip saw seems more like an extension of my hand. When I move my hand, the saw just seems to follow along."

Jigsaws come in two price ranges. Simple models, fine for occasional use, go for $35 to $70. Professional-grade tools start at $130 and are more likely to include the features Tom prefers: variable-speed control; a stout, tilting shoe; a quick-change blade system (no need to hunt for an Allen wrench or a screwdriver); and orbital action. Orbital action swings the blade slightly forward on the upstroke, helping the saw cut more aggressively. An orbiting blade also makes better progress rounding tight turns; it clears more of a path for itself with each stroke, so it's less likely to bind. (For a clean cut and reduced vibration when cutting metal, the orbit feature must be turned off.)

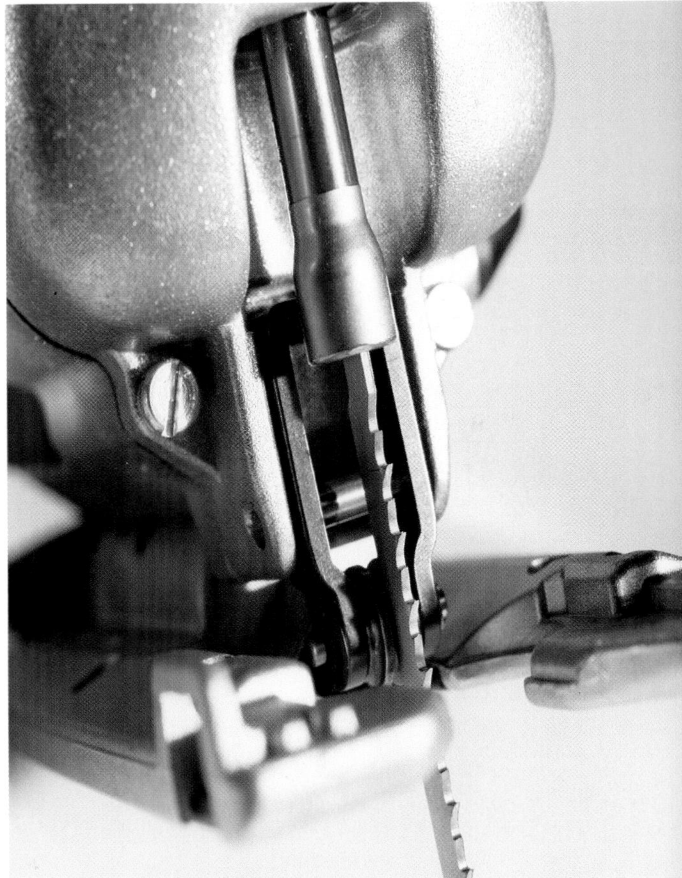

The teeth on jigsaw blades usually point up to help pull the tool toward the work on each upstroke. This also tends to spew sawdust over the cutline and splinter the surface. Most saws now have a built-in blower to keep the line clear, but when cutting something expensive like hardwood paneling, a few anti-splintering precautions should be taken. Some jigsaws come with a small plastic insert that fits into the shoe and surrounds the blade, helping to hold wood fibers in place. Tom often uses a metal blade to cut wood. "The tiny teeth reduce splintering. You can also score the cutline with a razor knife," he says, "as long as you remember to cut on the waste side of the line."

Other helpful techniques: Ease the saw more slowly into a cut, turn off the orbit feature, cut with the wood's good face down and swap a dull blade for a sharp one. Even then, what works on pine might mangle maple. To get the most out of a jigsaw, the trick is to hit on just the right combination of variables: blade aggressiveness, downward and forward pressure, degree of orbit and stroke rate.

JIGSAW FEATURES

1. Quick-change cap: Lift and turn to release the blade. Found on only a few saw models. **2. Variable-speed control:** Adjust as needed to get the smoothest cut. This one is trigger-mounted; on other saws, it may be near the motor or atop the handle. **3. Orbit control:** Look for several settings, from off to aggressive cutting. The smaller switch turns this saw's dust blower on and off. **4. Tilting base:** Allows the jigsaw to make beveled cuts. For scribing cuts, Tom Silva tilts the base a few degrees so the body of the saw leans toward the waste. The resulting undercut gives him a tighter fit. **5. Changeable shoe:** Cast aluminum adds stiffness; the steel insert can be swapped for plastic when cutting easily scratched materials.

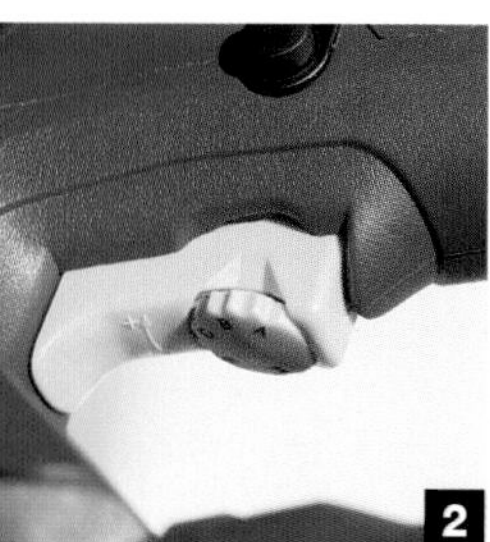

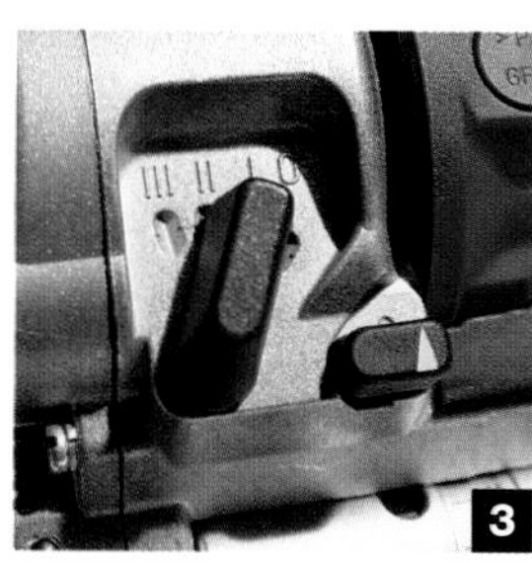

THE BLADES

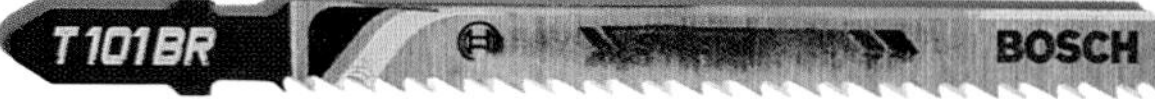

4-inch anti-splinter wood blade features downward-pointing teeth, bayonet shank. (10 teeth per inch)

Scroll-cut blade for wood. Hook shank. Teeth point toward shank, like those on most jigsaw blades. (10 tpi)

All purpose blade for wood. Universal shank requires screw. (6 tpi)

Medium carbide-grit blade for ceramic tile. Wide blade permits flush cuts.

3-inch blade for particleboard, plywood and similar sheet materials. (12 tpi)

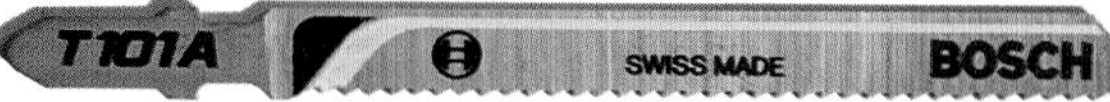

High-speed steel blade for medium metals and smooth cuts in wood. (14 tpi)

Thin-kerf metal-cutting blade. Good for cordless jigsaws, it uses less power than thicker blades. (24 tpi)

Bimetal blade (high-speed steel and high-carbon steel) for cutting stainless steel. (24 tpi)

Scroll-cut blade for smooth cuts and tight turns in wood. (20 tpi)

SHARPNESS

Tom Silva discards a blade the moment he suspects it's dull. "I gauge sharpness by the look of the teeth—sharp points and crisp edges." Check midblade, where most of the cutting happens. A dark patch on the blade means the blade has overheated and lost its temper. It will dull quickly, so toss it out before you lose your temper too. Make sure the replacement blade's shank is compatible with the saw.

SOURCES

Variable-speed orbital-action dustless jigsaw: barrel-grip, 1584DVS, $306; top-handle, 1587DVS, $317
Skil Bosch Power Tool Co. 4300 W. Peterson, Chicago, IL 60646; 800-815-8665.

JIGSAW BLADES

T101B2, 4" down-tooth antisplinter blade: 10 tpi, $9
Skil Bosch Power Tool Co.

Scroll-cut blade: #12379-J, bayonet-mount, 10 tpi, $9.25
Porter-Cable, 4825 Hwy. 45 North, Jackson, TN 38305; 800-321-9443.

All-purpose wood blade: #30012, 3 5/8", 6 tpi, $1.30
Vermont American Tool Co., Box 340, Lincolnton, NC 28093-0340; 800-742-3869.

Carbide-grit blade: RemGrit GJ18, 4" flush-cutting, $3.64
Greenfield Ind., Disston Div., Deerfield Industrial Park, S. Deerfield, MA 01373; 800-446-8890.

Blade for particleboard: #750022, 3", 12 tpi, $5.51
Hitachi Koki USA, 3950 Steve Reynolds Blvd., Norcross, GA 30093; 800-706-7337.

High-speed steel blade: T101A, 4", 14 tpi, $13
Skil Bosch Power Tool Co.

Thin-kerf metal-cutting blade: #792221-5, 24 tpi, $8.55
Makita USA, 14930 Northam St., La Mirada, CA 90638; 800-462-5482.

Bimetal blade: #750008, 3", 24 tpi, $12.47
Hitachi Koki USA.

Scroll-cut blade: T101AO, 3", 20 tpi, $5
Skil Bosch Power Tool Co.

Interchangeable-Tip Screwdriver: Model MR-502, $17-20
Latshaw Tools, 11020 Ambassador Dr., Suite 310, Kansas City, MO 64153; 800-833-3125.

Snappy Quick Chuck: #40011, $9.95
Poly-Tech Industries, 20610 Commerce Blvd., Rogers, MN 55374; 800-334-7472.

Screw Boss: #16062, $9.10
Vermont American Tool Co.

Vision: #784814A, $9.20
Makita USA.

CUTTING A HOLE

To cut the hole for a vanity sink, Tom draws his layout on the laminate and smooths masking tape over the lines to prevent the blade from chipping the surface. 1. After adding an extra cushion of tape beneath the leading edge of the tool's base, he fires up the saw and plunges its 4-inch, 14-tpi blade through the countertop. An alternative is to drill a pilot hole first. 2. Pressing down firmly to reduce chatter, he eases the saw forward, letting the blade, not his arm, do the work. 3. Tom stops several times to angle small nails into the edge of the cutout, which might otherwise rip loose as he nears the end of the cut, tearing the laminate and ruining his day.

1

2

3

screwdrivers

INTERCHANGEABLE-TIP SCREWDRIVERS

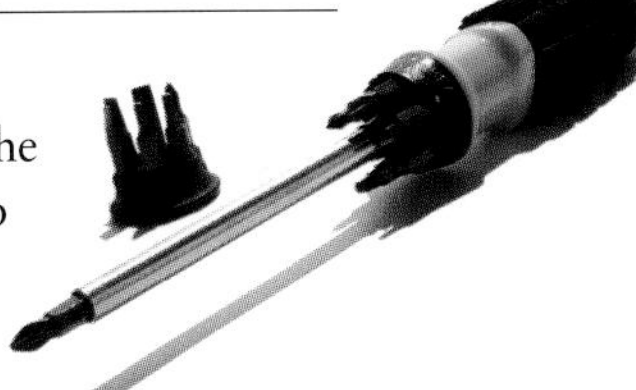

Steve Thomas thinks that interchangeable-tip screwdrivers make better sense than carrying around a bunch of screwdrivers on the job. We like the hefty, cushioned handles from Latshaw Tools. The top-of-the-line, $17 to $20 model shown here has a magnetized shaft and ratchet action.

QUICK-CHANGE SCREWDRIVER BITS

The Snappy Quick Chuck deserves its name: Chuck it into your 3/8-inch drill and you can install or change bits in seconds. Just push in a screwdriver, drill or counterbore bit, and it locks in place; to remove it, just slide back the locking collar. Problem: grooved hex-end tools are required. No problem: Snappy hex adapters fit round bits from 1/16- to 1/4-inch. Vermont American's Screw Boss is a handy tool to use with (or without) the Snappy. Its magnetic holder accepts hex-shank screwdriving bits, and its sliding sleeve holds screws up to No. 10 diameter. The sleeve keeps screws from tipping or wandering as the drill turns. Makita has a nearly identical accessory with a clear plastic sleeve so you can see when the screw is driven home.

sharpening

The best defense against sloppy woodworking is a sharp tool. Yet many woodworkers put up with dull ones because the sharpening process can be mysterious—and very frustrating when it doesn't turn out right. Toolmaker Leonard Lee's book, *The Complete Guide to Sharpening,* demystifies the task. Clearly a labor of love that took years to research, the book tells why, not just how, exploring everything from the various metals that go into tools to the crystalline structure of sharpening stones.

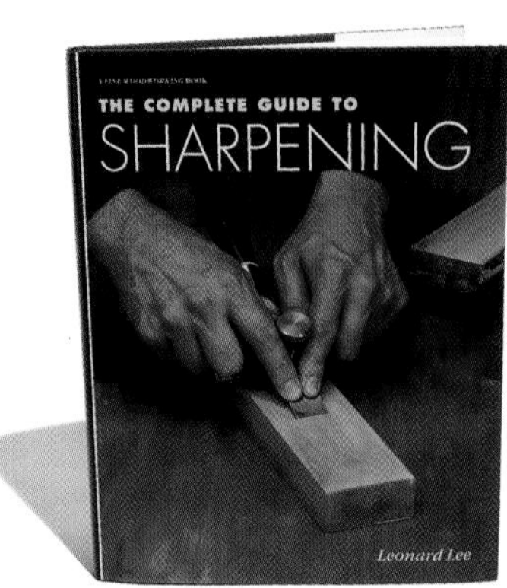

soldering

The secret to soldering is to have a clean joint with a perfectly round cross section. 1. First, measure and mark a length of tube. Rotate a tubing cutter around the tube, tightening the handle slightly at each turn to avoid flattening the tube. 2. Remove the burr inside with the triangular reamer that folds out of the cutter; keep both edges of the reamer in contact with the tube so the tube stays round, and don't push too hard or the end will flare. 3. Clean the inside of the fitting with a fitting brush, turning it clockwise to avoid breaking the wire bristles. Use a brush the same diameter as the tubing; don't touch the surface afterward. 4. Polish the outside of the tube with fine-grit emery cloth. 5. With a disposable acid brush, coat both mating surfaces with a thin layer of flux to neutralize oxides, then insert the tube into the fitting. Clean, flux and fit all pieces together. 6. Uncoil about 7 inches of solder wire; bend a 2-inch crook near the end. Holding a propane torch at an angle to the fitting, train the innermost blue flame on the fitting. When the flux bubbles, touch the tip of the solder wire to the side of the joint opposite the flame. (Never put flame on the solder itself.) When the solder liquefies and is sucked into the fitting, remove the flame and run the tip of the solder around the joint until it is filled. Gently wipe the joint with a wet cloth to set the solder, then wait until all parts have cooled.

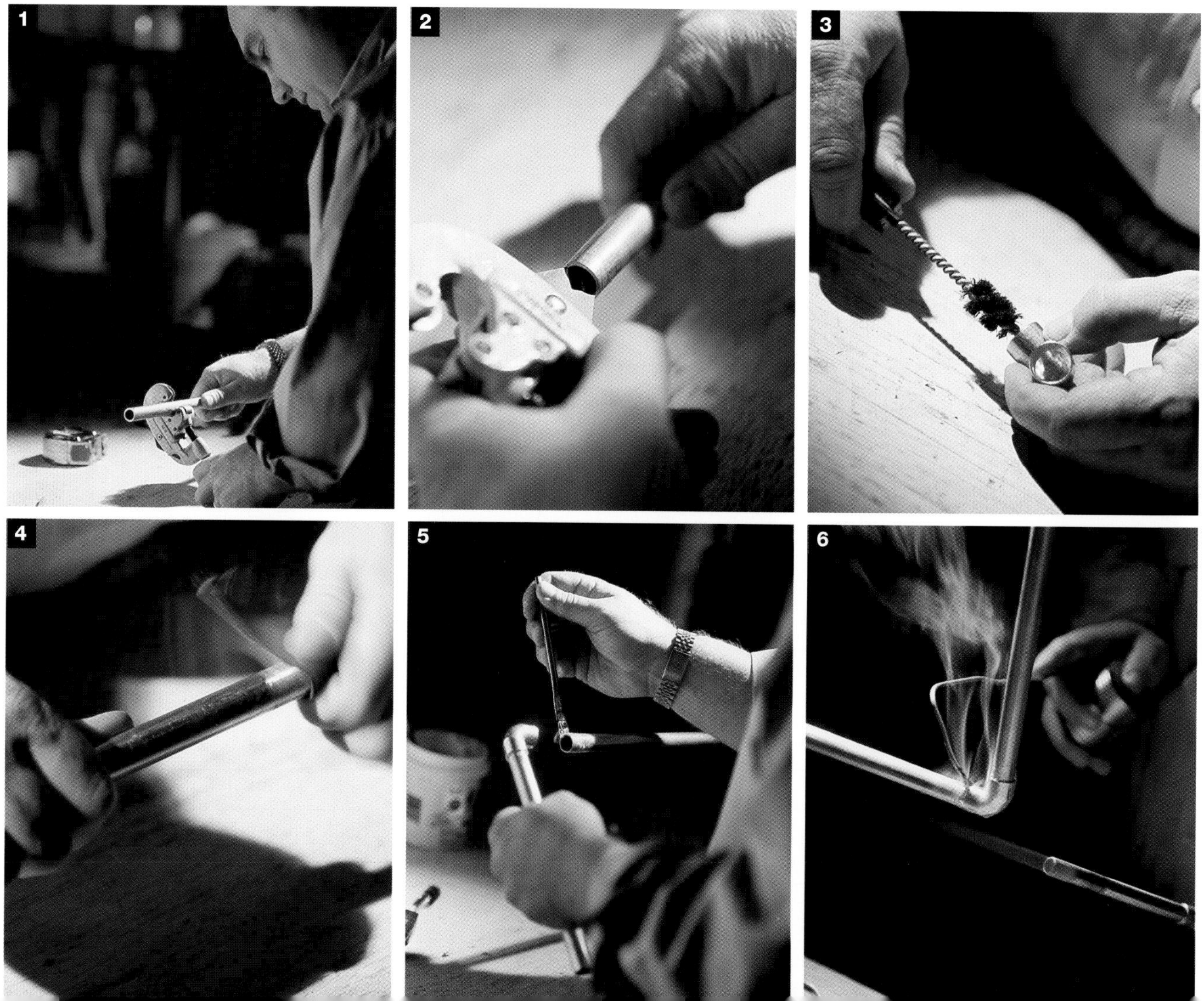

1

2

3

4

5

6

The Complete Guide to Sharpening, by Leonard Lee, 1995, 240 pp., $34.95
The Taunton Press, 63 South Main St., Box 5506, Newtown, CT 06470-5506; 800-888-8286.

Propane torch: Model TS 2000K two-piece, trigger-start kit $31.56
BernzOmatic, 1 BernzOmatic Dr., Medina, NY 14103; 716-798-4949.

Solder: 100% Watersafe, 1-lb. spool, $6.80
M.C. Canfield & Son., 1000 Brighton St., Box 3100, Union, NJ 07083; 908-688-5050.

Flux: Nokorode Soldering Paste, 1.7-oz. can, $1.29
The M.W. Dutton Co., 3 Bridal Ave., Box 232, West Warwick, RI 02839; 401-821-1832.

Fitting brush: Model 6100, $2
Mill-Rose Co., 7995 Tyler Blvd., Mentor, OH 44060; 216-255-9171.

Emery cloth: 1½" x 10–yard roll, $5.95
Durst Corp., 115 Globe Ave., Mountainside, NJ 07092; 908-789-2880.

Fittings
Nibco, 500 Simpson Ave., Elkhart, IN 46516; 219-295-3000
Elkhart Products Corp., Box 1008, Elkhart, IN 46515; 219-264-3181.

Tubing cutter: #125, $13.75
General Hardware Mfg. Co., 80 White St., New York, NY 10013; 212-431-6100.

Instant patina solution: Patina Green, 8-oz. bottle, $11.45
Constantine's, 2050 Eastchester Road, Bronx, NY 10461; 800-223-8087.

FURTHER READING

Plumbing a House by Peter Hemp, 1994, 224 pp., $29.95
The Taunton Press, 63 S. Main St., Newtown, CT 06470; 203-426-8171.

The Copper Tube Handbook, 1995, 50 pp., $4
Copper Development Association, 260 Madison Ave., New York, NY 10016; 212-251-7200.

TIP

Soldering Safety

- Protect nearby materials with a woven-fiber heat shield, available from plumbing supply stores.
- Water and hot solder don't mix. If the inside of the tube is wet, stuff a wad of white bread (no crust) into it. The bread absorbs drips while you're soldering, then dissolves when the water is turned on.
- Flux is caustic: don't get it on your skin or in your eyes. Wipe all residue off joints (it may corrode the copper) and flush plumbing before drinking water.

tool belts

Durability is the first thing to look for when shopping for a tool belt. Tightly stitched seams and rust-resistant rivets are a must, and the pouches should be made from strong material. Canvas or cotton may be good enough for clerks at the lumberyard, but carpenters need something tougher to resist wear and tear from carrying tools.

Leather is the traditional material. Tool belts made of top-grain cowhide—smooth on the outside and rough on the inside—take a lot of punishment; they also take a long time to break in. Many carpenters would need a lot of convincing before they bought a belt made of anything but thick cowhide. On the other hand, some carpenters opt for suede, which is softer and easier to break in. The downside is that it's thinner and may wear out sooner.

Nylon is a less expensive alternative to leather. Cordura, an abrasion-resistant nylon fabric, is light and won't mildew or crack if it gets wet. Jose Munoz, product services manager for tool belt maker McGuire-Nicholas, says a Cordura belt should last twice as long as a good leather one. "The biggest drawback is that nylon won't conform to the body," he says. "It's like wearing plastic shoes instead of leather ones."

There are even more options when it comes to choosing the belt's harness: narrow or wide, metal buckles or plastic clips, front or back release. Adjustability is the key, especially if the belt will be worn over light clothes in the summer and bulky wraps in the winter. *This Old House* contractor Tom Silva grew tired of belts that can only be tightened or loosened in discrete increments, so he came up with a novel solution. He snipped off the buckle on his tool belt and replaced it with one from a car seat belt.

Size and number of pockets are largely matters of taste. Some carpenters prefer pouches that slide along the belt. Others like pouches that are sewn in place so they won't slip. In either case, the pockets should have openings large enough to reach into and smooth seams along the edges instead of the bottom, where they can trap nails or other odds and ends. A good tool belt will last at least three or four years, perhaps longer if it is restitched.

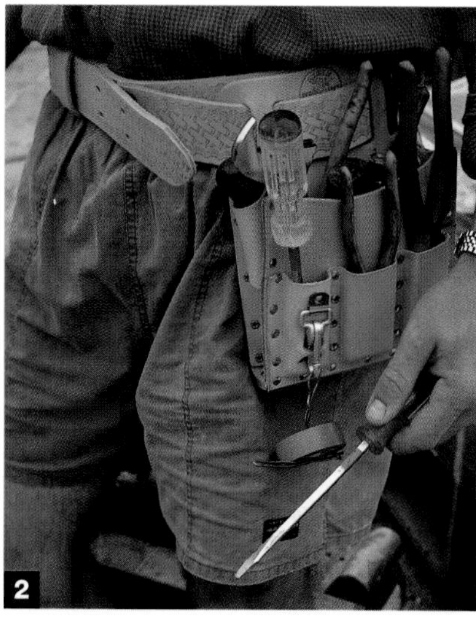

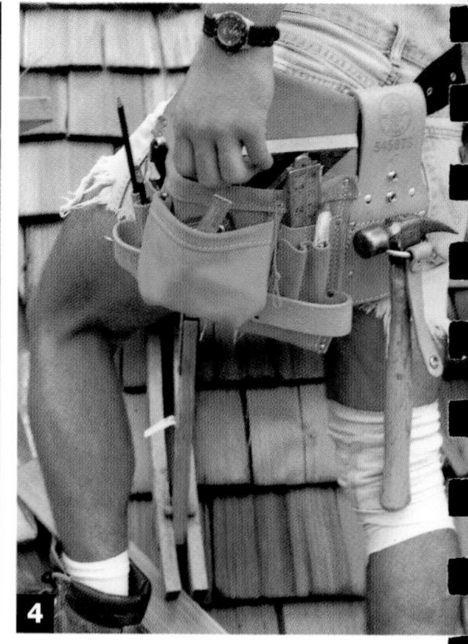

1. Ideal for framers, this colorful Cordura model has 10 capacious pockets, which can easily swallow up a nail gun or a drill. Nice touches include a back pocket with a Velcro flap—ideal for stashing valuables.
2. Electricians use fewer tools than carpenters, so their belts have pockets on only one side. This thick cowhide model has plenty of room to store wire strippers, screwdrivers, snips and needle-nose pliers. The T-shaped bar is for holding rolls of tape; the hook is for hanging a voltage meter.
3. A leather cage at the front of this soft suede belt keeps a measuring tape handy, and the pockets are deep enough to accommodate a finish carpenter's nailsets, putty knife, glue, snips, block plane, screws, sandpaper and lots of pencils.
4. For roofing work, a tool belt with a single pouch that won't get in the way is a good choice. But if there isn't enough space for all the nails, this tool belt can be customized by adding more pouches. One unique accessory: a metal hammer loop with a safety strap.

NORM'S SIDEKICK

Norm Abram is so attached to his cowhide belt that he wears it around the shop instead of grabbing tools off the shelf. "It's so easy to have all you need right at your waist," he says. This belt, which has been his faithful hipside companion for years, has a cotton-web harness with a plastic buckle in back and is very easy to put on and take off. The honey-colored pouches, with darkened edges polished by age, have developed the satisfying sags and bulges that make a belt comfortable and give it character. "I like the way the leather feels," Norm says. "Smooth and substantial."

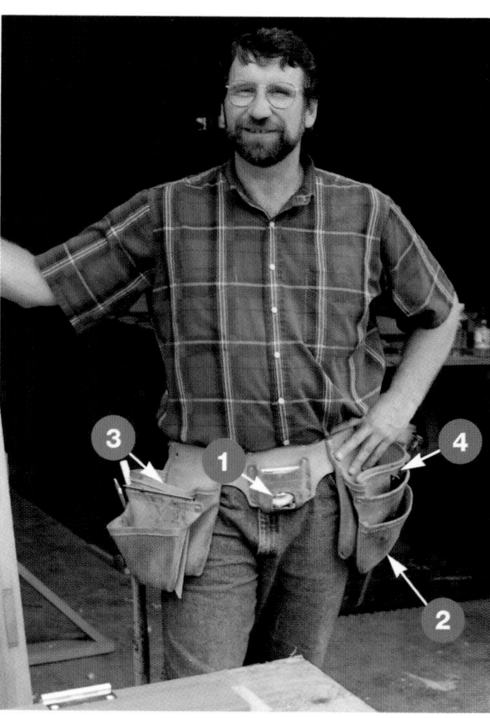

1. Norm's tape measure has a home in the front of the belt. 2. He reserves the shallow outer pockets for nails, screws and sandpaper and tucks bigger items in the deep inside pockets. 3. He fills the inside right pouch with countersinks, driver bits, a compass, a wood rasp and even a pair of tweezers. 4. When the job requires more tools, he stashes them in the left inside pouch.

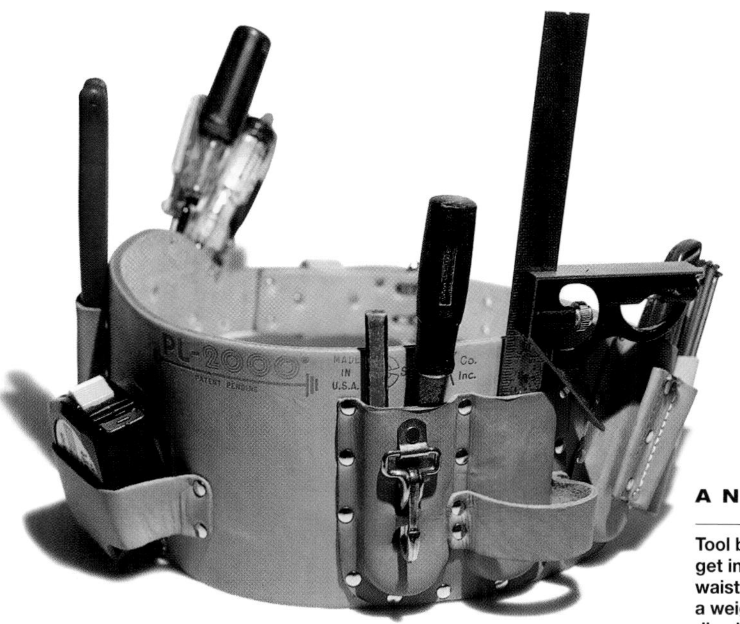

AN ALTERNATIVE

Tool belts that hang from the waist can be uncomfortable, spill tools when you sit and get in the way of a nail apron. Seatek's new Super-Belt suspends the tools above your waist and around your back and sides, apportioning the weight more evenly. Modeled on a weightlifter's belt, it also provides back support. Despite its comfort, we found some disadvantages. For one, reaching around to fish tools out of the back pockets takes a little getting used to. For another, you can't safely sit on upholstered chairs.

Framer's belt: Deluxe set #1C-55, $130
Diamond Back USA, Box 347179, 1121 Silliman St., San Francisco, CA 94134; 800-899-2358.

Electrician's belt: Tie-wire leather belt, #5420, 2" wide, 36" to 44" waist, $27; 6-pocket leather, #5127, fits belts up to 2" wide (models available to fit belts up to 2½"), $24
Klein Tools, Inc., 7200 McCormick Blvd., Box 599033, Chicago, IL 60659-9033; 847-677-9500.

Finish carpenter's belt: Leather Jural B-283, $47
Garrett Wade Co., 161 Ave. of the Americas, New York, NY 10013; 800-221-2942.

Nylon-web suspenders: #60209R (red) with adjustable leather back-spreader, $11
Klein Tools Inc.

Roofer's belt: Polypropylene web belt #5225, 2¼" wide, adjusts from 32" to 48" waist, $7; 6-pocket leather nail, screw and tool holder, #42241, fits belts up to 3" wide, $22; Leather hammer holder, #5456TS, fits belts up to 4" wide, $11
Klein Tools Inc.

Super-Belt: large $74.95; X-large, $74.95; XX-large, $78.25
Seatek, 392 Pacific St., Stamford, CT 06902; 203-324-0067.

Contractor Grade Dynagrip Retractable Knife: #10-779, $8
Stanley Tools, 600 Myrtle St., New Britain, CT 06053; 800-262-2161.

Super Heavy Duty Retractable Blade: #10-122, $15
Stanley Tools.

Magazine Load Snap-Off Blade Utility Knife: #42048, $9.98
Hyde Tools, 54 Eastford Rd., Southbridge, MA 01550; 800-872-4933.

PowerGrip Utility Knife: #7845, $10.79
Fiskars Inc., Wallace Div., 780 Carolina St., Sauk City, WI 53583; 800-500-4849.

199 Contractor Grade Fixed Blade Knife: #10-209, $6.50
Stanley Tools.

Retractable Banana Knife: #60106, $19
Better Tools, Inc., 3030 Centre Pointe Dr., Suite 400, Roseville, MN 55113; 800-798-6657.

A PERSONAL PORTABLE TOOL BOX

There is no perfect tool belt. Subtle differences—in the width of the harness, the way the belt buckles and the size and number of pockets—may meet the needs of one workman but not another. Some manufacturers offer special slide-on pouches and tool-holding accessories that can be matched to a particular type of task. If a pouch wears out, it can be tossed without sacrificing the whole belt.

utility knives

Over the years, utility knives have acquired a degree of sophistication far beyond their humble beginnings. There are knives with blade-storing handles, tool-free blade-changing, brightly colored "designer" casings and cocked handgrips for comfort and leverage. The latest versions have locking mechanisms that keep retractable blades from wobbling. Despite the advances, plenty of renovators clamor for the old classic, Stanley Tools' knife No. 199, which is still manufactured but now called Model 10-209.

1. **A new British-made model is so large (7½ inches long), it comes with a molded plastic holster. Center button tightens blade to hold it firm while cutting.**
2. **Blades on the Hyde 42048 snap off along a scored edge when they become dull.**
3. **Rubber grips on Stanley Model 10-779 make the knife easier to hold. A screwdriver is required when changing blades.**
4. **Fiskars' PowerGrip has a protective knuckle-duster grip and an easy-access storage compartment that holds three replacement blades.**
5. **The original utility knife, No. 199 is virtually identical to the 1937 version—compact and simple.**
6. **This Better Tools banana knife is named for its color and shape. The 19-degree bend in the handle means more power in each cut. Grip swings open for no-tool blade changes.**

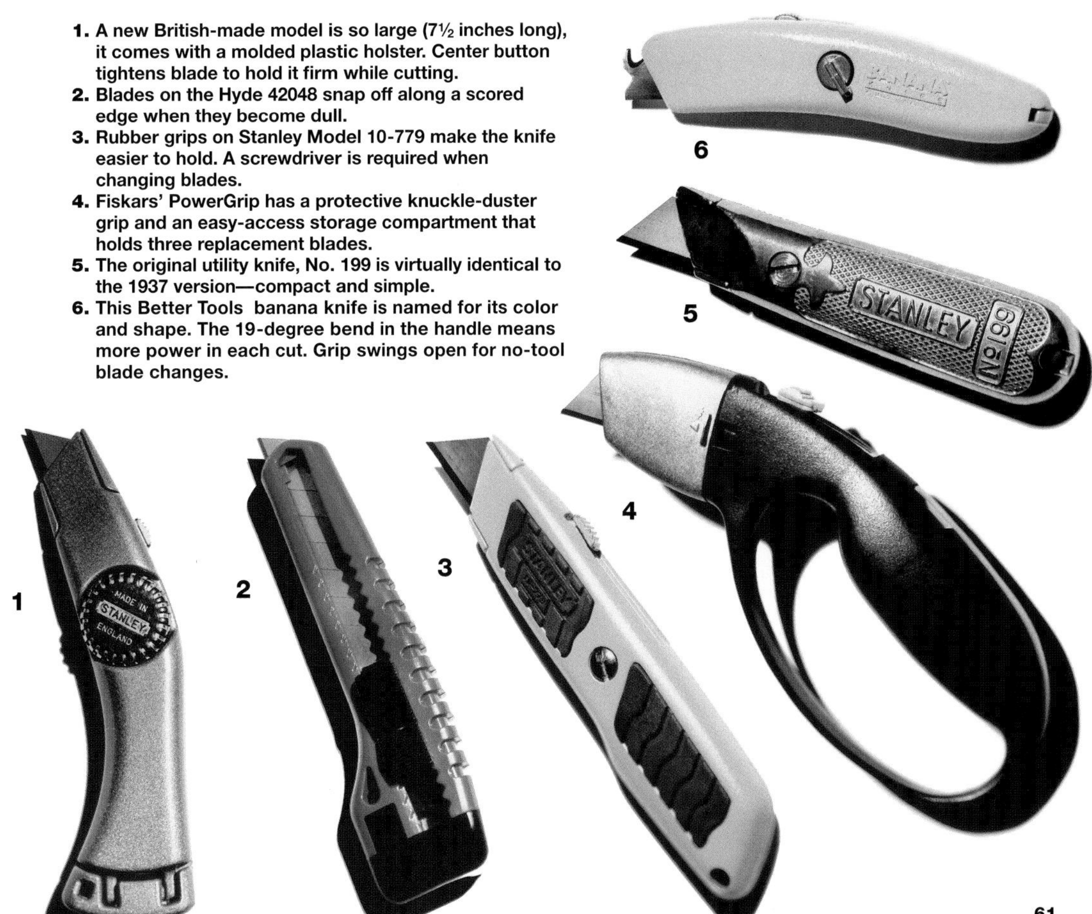

workbenches

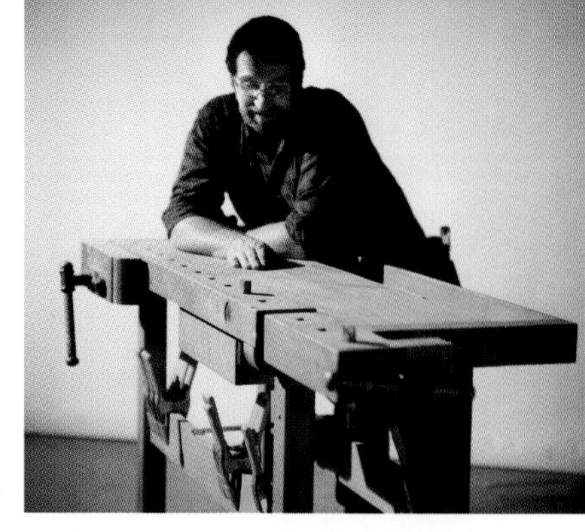

WORKBENCH FEATURES

THE ACCESSORIES

There are hundreds of bench accessories, only some of which are really useful. Some woodworkers swear by specialized stops and jigs. Norm gets best results with off-the-shelf clamps. Here, he uses them to secure a long section of a tabletop while cutting slots with a biscuit joiner. The weight of the bench keeps it from hopping across the room as he makes the cuts.

THE BENCH DOGS

Bench dogs are movable stops that, paired with vises, secure projects to the bench surface. Which type of dog works best is debated endlessly among woodworkers. Metal dogs don't break like wooden ones, but they mar the edges of stock and are more expensive. Round dogs permit clamping from any angle, but square ones offer better holding power. And so on.

THE BENCH TOP

A well-made bench is a beautiful piece of furniture. Some woodworkers spend more time sanding and oiling the bench top than working on it. A few cuts from a stray blade give character, but Norm uses a sheet of cellulose fiberboard to protect the top during messy work, like gluing up projects.

THE TAIL VISE

Vises keep the wood from moving while you work it. Without them, the workbench is little more than a glorified table. Pieces of stock are typically wedged between bench dogs located on the tail vise, which opens off the end of the bench, and the bench top.

THE WORKSTATION

The workbench is a tool, but it is also a workstation for the router, the mini-lathe or, in this case, an orbital sander. Norm uses a section of rubber mesh (like the mats that keep area rugs from slipping) as a router pad and, here, to anchor the project that's being sanded.

THE FACE VISE

Another debate among woodworkers is over what type of vise works best: wood or metal. Wood jaws, shown here on the face vise (on the side of the bench), won't leave marks on the project being clamped. Wood screw mechanisms are traditional, but metal is superior because it won't fatigue, Norm says.

SOURCES

Build-It-Yourself Workbench: #NY1021, plans only, $10; #NY1022, video and plans, $24.95 plus shipping and handling (approx. $125 for parts)
Box 9345, S. Burlington, VT 05407-9345; 800-892-0110.

Veritas bench: $795, plans for $9.95, basic bench kit for $49.95, deluxe bench kit for $299
Lee Valley Tools, 12 E. River St., Ogdensburg, NY 13669; 800-871-8158.

Sjobergs bench: $390, available through several catalogs, including
Trend-Lines Inc., 135 American Legion Hwy., Revere, MA 02151; 800-767-9999.

Workmate 200: $124
Black & Decker, 701 E. Joppa Rd., Towson, MD 21286; 800-762-6672.

Shop Boss: $59.95 plus shipping
PMI, 3655 East Roesar Road, Phoenix, AZ 85040; 800-325-6952.

WHAT TO LOOK FOR

To withstand all the pounding, pummeling, pushing and pulling that's done on it, a workbench must be strong. The base should be broad, with trestles or braces to keep the bench from racking. The joints—mortise and tenon are best—should be tight. Most important, the top should be made from a close-grained wood, such as maple, birch or beech, to add weight (Norm's bench is a hefty 300 pounds) and keep the bench from skittering and bouncing while you're planing a board or cutting dovetails. A dense wood is more dimensionally stable than a soft one; the top won't shift around as much with changes in humidity. And a good, thick top can be planed if it's marred.

Whether you plan to buy a bench or build your own, try one on for size first. The bench should suit your working habits and maximize your efficiency. It should be tall enough to prevent backaches and low enough to provide leverage. Select the largest workbench your space can hold. Like a desk or the kitchen table, benches have a way of filling up with stuff quickly.

Someday you'll find the time to design and build your own perfect bench. We know that. Until then, here are a few ready-built models and plans to choose from.

BUILD-IT-YOURSELF WORKBENCH

Building your own bench is a good first project for budding woodworkers. This one incorporates several different types of joinery, including dadoes and rabbets. You can add vises and other accessories as you need them.

VERITAS

This Canadian-built model, available in kits or fully assembled, is good enough for woodworkers who make a living off their bench. It features a maple top, a twin-screw vise and a sturdy trestle base. Total weight: 200 pounds.

SJOBERGS

Weighing in at 125 pounds, this Swedish-built, birch-topped bench isn't as beefy as the professional models, but it's fine for the semiserious woodworker. It comes with two vises, plastic dogs and plenty of storage space in the base.

BLACK & DECKER WORKMATE 200

Built for site work, the top of the portable Workmate opens to form a vise or closes to create a work surface. The collapsible steel frame includes a footstep that you use to give the bench stability and weight.

A BETTER SAWHORSE

When you need something steadier than a sawhorse and bigger than a Workmate but as portable as both, the Shop Boss fits the bill. Built of 18-gauge steel and 2x3 lumber, it weighs 25 pounds, stands 31 inches high and provides plenty of work space: about 62 by 30 inches as a skeleton, lots more with a sheet of plywood laid on top. Tap the locking cams and the legs fold, so you can hang it on a wall, stow it behind a door or lug it outside as a base for a picnic table.

Robo-Grip, RG9 curved jaw or straight jaw, $29.95; RG6 straight jaw or curved jaw, $27.95
Applied Concepts Inc., 120 Marshall Dr., Warrendale, PA 15086; 412-776-5595.

Knipex #KN8603, (manufacturer lists it as 250 mm., we measured it as 10½"), $73.50
Knipex, Anglo American Ent. Corp., 403 Kennedy Blvd., Somerdale, NJ 08083; 800-223-8600.

#KN8711-10, (manufacturer lists it as 250 mm., we measured it as 9¾"), $35.60
Knipex.

#KN8711-7, (manufacturer's listing 7"), $31.70
Knipex.

Sunico EZ-Grip, $17.99
Sunico Inc., 13727 Excelsior Dr., Santa Fe Springs, CA 90670; 800-483-1788.

wrenches and pliers

We're always on the lookout for tools that make gripping easier. The Robo-Grip seems like a gizmo at first, but it's simple to use. It self-adjusts as you squeeze the handles, which are well shaped. Of riveted, laminated construction, it's also light, and the toothed jaws lock parallel on pipes and nuts. It's available in 9-inch and 6-inch models, with curved or straight jaws. The spring-loaded handles tend to stay open in the toolbox, however, so it can get tangled with other tools. Two Knipex imports also work well. Although it doesn't self-adjust and its handles can pinch your palm, Model KN8603 (10½ inches) has smooth, parallel jaws that are good on nuts. The lighter model KN8711-10 (9¾ inches) is a lot handier. The handles are springy and comfortable, and the jaws have two bites, one for pipe and another for nuts. As for Sunico's EZ-Grip, an 8-inch adjustable wrench with a handy thumb-slide mechanism, it's not heavy-duty enough for hard use, but it's a good occasional tool for quick fixes.

Robo-Grip RG9 curved jaw

Knipex KN8603

Knipex KN8711-10

Sunico EZ-Grip

notes

wood products

BUYING LUMBER | PLYWOOD | TREATED WOOD

FINIALS | WOOD FLOOR FINISHES

buying lumber

Lumber doesn't come shrink-wrapped with instructions. Buying the best lumber for a job means buying lumber best suited to that job, not necessarily the highest grade or the prettiest. Lumber is graded according to how it should be used—the highest grades for finish work, lower grades for different levels of construction.

The do-it-yourselfer, especially in a home-center environment, is tempted to buy what looks the best, but in reality he or she should buy the grade that's appropriate for the job. Don't buy No. 1 clear pine for trim, especially when you're going to paint it. Buy No. 2 and save yourself some money.

Besides understanding something about grades, the next most important thing about building economically is buying the right *amount* of lumber. Plan out your project to scale on a piece of graph paper. Figure out how best to assemble the pieces and how long they need to be. Change your project dimensions, if necessary, to accommodate standard lumber sizes. When you are satisfied with the layout, count up the pieces of each length you need. Some may be short enough to be cut from longer lengths with less net waste.

Many would rather buy lumber from a lumberyard than a home center. Home centers carry lumber, but it's not their primary business. They don't carry a broad selection. Lumberyards have beams, engineered lumber, high-quality clapboards, moldings—thousands and thousands of feet of them—and all thicknesses of plywood. Besides, the quality of the lumber in a home center is often not as good as that in a lumberyard. Another clear advantage to shopping at a lumberyard is that you are more likely to find an expert you can trust. A manager at a lumberyard assumes that a do-it-yourselfer does not necessarily understand grades and can help select the stock for a project.

You don't want to overbuy, and you don't want to make many other mistakes. It's very possible to buy some lumber that won't be in stock in the same size a month later when you come back to pick up more so you can finish the project. Experienced lumberyard managers will help you avoid making such potentially costly errors.

LUMBER BASICS

Lumber comes from trees, at least initially, and trees are either hardwoods or softwoods. Hardwoods are deciduous trees, which lose their leaves each year, grow slowly, take up a lot of space in the forest and don't naturally grow very straight; therefore, they are the more expensive woods. They are beautiful and densely grained woods that take carving, shaping and finishes well. Maple, cherry, birch, walnut and oak are the most common.

Softwoods come from the needled conifers and grow quickly, straight and tall; they are therefore more economical to forest and mill. They constitute the bulk of construction lumber and a high percentage of the components of engineered woods. Pine, fir, hemlock and spruce dominate the softwoods. Redwood and cedar are also softwoods, but they grow more slowly, are more expensive and are generally sold for exterior use—claps, fences, decks—due to their high rot resistance.

Lumber is sold in three broad categories: *construction lumber*,

pattern lumber and *specialty products*. Construction lumber is separated by size into three categories: *timbers*, which are the biggest pieces, over 5 inches wide, for structural purposes; *dimensioned lumber*, which is between 2 inches and 5 inches wide; and *boards*, anything under 2 inches thick. The cheaper grades of boards are used for fences, sheathing and shelves, the better grade for finish carpentry. Most construction lumber is softwood.

Pattern lumber is milled into particular shapes, such as clapboards, moldings and flooring. It can be either hardwood or softwood.

Specialty products include hardwoods and hardwood veneers, as well as all lumber products that have been processed, treated or engineered to one degree or another—plywood, laminated beams, I-beams, oriented strand board or pressure-treated wood.

LUMBER GRADING

After it has been milled, lumber is graded. Grade is calculated according to a complex formula that considers the type, size, closeness, frequency and location of all characteristics and imperfections affecting appearance, strength and use. Associations established for the regulation and standardization of lumber grading assure consistency for all lumber products and by-products. Hardwood lumber grades are based primarily on the presence or absence of natural or manufacturing defects that affect the appearance of the wood. Softwood lumber grades emphasize strength rather than looks.

On the following page are some thumbnail facts about lumber grades. Keep in mind that most lumber for do-it-yourself projects will fall in the high end of graded lumber, above No. 2.

A 2X4 IS NOT REALLY 2X4

Why is this? Unlike length, the thickness of wood is commonly measured in "nominal" dimensions, which are greater than the actual dimensions due to surfacing, drying and parsimonious lumber manufacturers. Therefore, a board might start out as 2 inches and end up being 1½, or 4 inches ends up as 3½. There are limits, however, to the differences: The American Softwood Lumber Standard (ALS) specifies that a 2x4, for instance, must measure at least 1½ by 3½ to legitimately bear that name.

WHAT IS A BOARD FOOT?

A board foot is a standard unit of volume based on a piece of wood 12 inches by 12 inches with a thickness of 1 inch before it is dried. That means a board foot of 1x4 is 3 feet long; a board foot of 2x12 is 6 inches, or half a foot. A formula for figuring out board feet looks like this:

$$\frac{\text{thickness (inches)} \times \text{width (inches)} \times \text{length (feet)}}{12} = 1 \text{ board foot}$$

SOFTWOOD LUMBER GRADES

Finish Product Grades

B & BETTER	Highest recognized grade of finish. Generally clear, although a limited number of pin knots are permitted. Finest quality for natural or stain finish.
C	Excellent for painted or natural finish where requirements are less exacting. Reasonably clear but permits limited number of surface checks and small tight knots.
C & BETTER	Combination of B&Better and C grades; satisfies requirements for high-quality finish.
D	Economical, serviceable grade. Has minor defects on one side but larger defects on the other side. Often used for interior woodwork that will be painted.

Common Grades for Boards, Dimensional Lumber and Timbers

SELECT STRUCTURAL	High quality, relatively free of characteristics that impair strength or stiffness. Moderate to good appearance.
No. 1 **No. 1 DENSE** **No. 1 NONDENSE**	For general utility and construction. Where high strength, stiffness and good appearance are desired.
No. 2 **No. 2 DENSE** **No. 2 NONDENSE**	For most general construction uses. Allows well-spaced knots of any quality.
No. 3	For general construction, appearance not a factor.
STUD	Comes in 2-inch to 6-inch widths only. Suitable for load-bearing walls.
CONSTRUCTION	2-inch to 4-inch widths only. Good for general construction, graded primarily for strength and serviceability.
STANDARD	2-inch to 4-inch widths only. Similar to construction grade, but characteristics are limited to provide good strength and excellent serviceability.
UTILITY	Recommended where a combination of economy and strength is desired. Used for such purposes as studding, blocking plates, bracing and rafters.
ECONOMY	Usable lengths suitable for bracing, blocking, bulkheading and other utility purposes where strength and appearance are not important.

HARDWOOD LUMBER GRADES

FIRST AND SECONDS (FAS)	The FAS grade consists of long, wide, clear cuttings best suited for high-quality furniture, interior trim and solid wood moldings.
FAS1FACE OR SELECT S	This grade is determined using both faces of the board. The best face must meet the requirement for FAS and the reverse must meet the standards for No. 1 Common.
No. 1 COMMON	Clear cuttings of medium length and width. Best suited for furniture, cabinets and a multitude of solid-wood manufactured products.
No. 2 COMMON	Short, narrow, clear cuttings economically priced. Ideal for use in unexposed furniture frames, picture frames, cabinet rails and frames and novelties.

Above material courtesy Georgia-Pacific

TOM SAYS

- Leave a ⅛-inch space between wall, roof and subfloor panels to allow for expansion and contraction. Special H-clips or 6d nails will give the right spacing.
- Lay plywood so its long dimension (its strongest) is perpendicular to joists, rafters or studs.
- Stagger joints. Where four corners meet, overall strength suffers.
- The more plys a panel has for a given thickness, the greater its resistance to warping.
- For roof sheathing, use ⅝-inch tongue-and-groove ply over rafters 24 inches on center.
- For subfloors, use ¾-inch ply on joists 16 inches on center.
- For walls, use ½-inch ply over studs 16 inches on center.

EXTERIOR PLYWOOD
The workhorse of laminated panels is made with softwoods and water-resistant glue. The wood decays if kept in contact with soil or water. In such situations, pressure-treated panels last longer.

plywood

Plywood has never entirely shaken its image as being inferior to real wood, even though plywood panels cover roofs, hold up floors and stiffen walls in the majority of homes in the United States. Ironically, because so many builders now use even cheaper wood-chip-and-glue panels, plywood-sheathed houses are seen as a sign of higher-quality construction.

Composed of wood sheets so delicate a child could easily snap one in her hand and held together by inexpensive plastic glues, the assembled piece of plywood is stronger, stiffer and less prone to warping than solid wood of the same thickness. The secret is gluing the veneers so that each grain is perpendicular to the next. The resulting "wood" is so tough that a half-inch sheet can support a compact car.

IS PLYWOOD DANGEROUS?

Nobody should inhale formaldehyde, a stinky chemical for preserving biology-class frogs. Yet in today's tighter, energy-efficient buildings, people regularly breathe formaldehyde gas given off by carpeting, paneling and plywood. Minute quantities, on the order of 0.005 parts per million, can trigger sensory irritation; higher concentrations can cause watery eyes, nausea and asthma.

The urea-formaldehyde glues commonly used in interior plywood (as well as particleboard and fiberboard) release formaldehyde in amounts that reach the limits set by the U.S. Department of Housing and Urban Development: 0.2 parts per million for plywood and 0.3 for particleboard. By contrast, exterior-grade plywood and strand board panels, made with phenol-formaldehyde resins, emit such insignificant amounts of gas (about 0.1 parts per million) that HUD exempts them from testing. Formaldehyde emissions decrease with time, but in poorly ventilated houses or among highly sensitive individuals, even short-term exposure is undesirable. To avoid breathing the gas, use solid lumber, exterior-grade panels or interior plywood with a waferboard core.

PLYWOOD LABELING

All plywood is graded A through D based on defects in its face and back veneers. When contractors spec ply, they match grade to purpose. For hidden wood such as sheathing or subflooring, *This Old House* contractor Tom Silva uses cheaper CDX ply, which has a C-grade face, D-grade back and glue rated for Exposure One. When looks count, as in cabinetwork, he'll pay more for A-grade ply, which has fewer knots and fewer of those football-shaped patches called plugs or boats. But for structural panels, performance, not appearance, is what counts. The U.S. Department of Commerce developed a voluntary performance standard in the late 1960s. Today, most sheathing-grade plywood and strand board bears the stamps of testing agencies like the American Plywood Association, Teco or Pittsburgh Testing Laboratories, which rate the panel's suitability for a particular application, such as roofing or subflooring, regardless of grade.

SUBSTITUTE PLYWOOD

Oriented strand board, referred to as OSB by contractors, is increasingly replacing plywood in construction because it is as strong as plywood, costs less and has fewer voids. It is made from layers of wafers that are shaved off logs, roughly aligned at 90 degrees to each other and hot-pressed with phenolic resin glue. The manufacturing process incorporates 90 percent of a tree and uses young, fast-growing trees such as aspen, poplar and pine. By contrast, plywood manufacturers use about 60 percent of a tree and rely on logs 60 to 70 years old.

Introduced 15 years ago by Potlatch Inc., strand board has captured more than a third of the sheathing market, but not everyone likes the stuff. "I don't use it unless I have to," says Tom Silva. "It's heavy and slippery, a real danger on pitched roofs. And it does a job on my saw blades." (The high glue content is not friendly to tools.) "If it gets wet, it'll swell so much you'll trip over it, or the joints show through the shingles." Although manufacturers seal edges, they recommend keeping panels dry at all times. Says Paul Fisette, program director of building materials technology at the University of Massachusetts: "Plywood had its problems too at the outset. Strand board technology will improve."

SPECIALTY PLYWOODS

Almost anything can be put between two sheets of wood. Innovative plywood cores improve stiffness, reduce weight and create products for specific uses, all for a premium price.

AIRCRAFT

Favored by hobbyists who need lightweight skins for their model aircraft. This sample is 1.5 mm thick and has three laminations of okoume (an African mahogany), each just 0.5 mm thick.

FLEXIBLE

For cabinetmakers who want a curve instead of a corner, these plys can be bent like a sheet of rubber. Thickness and number of laminations determine the flexibility of the finished product.

MARINE

The ultimate water-resistant panel, often made with special boil-proof glues and defect-free tropical hardwoods. The best are registered with Lloyds of London and start at about $150 for a ¾-inch-thick 4 by 8 sheet.

MEDIUM DENSITY OVERLAY

Paper impregnated with phenolic resin is bonded to a plywood core. Used on cabinets, soffits, exterior trim and where a smooth, paintable surface is required.

BALSA CORE

A lightweight panel filled with blocks of balsa. Used for bus and aircraft floors and the interiors of boats.

PAPER CORE

This honeycomb paper core weighs next to nothing and when laminated to veneer makes an incredibly stiff panel. Used in exhibit-booth floors, boat and aircraft interiors or anywhere lightness is critical.

INTERIOR PLYWOODS

Woodworkers favor these smooth, hardwood-faced panels for making cabinets and bookcases. The glues won't tolerate water and may release significant amounts of formaldehyde gas.

VENEER CORE

A favorite of cabinetmakers. Made with solid hardwood veneers throughout. European-made versions are known as Baltic birch.

HYBRID CORE

Particleboard cores provide a smooth, stable substrate for the face veneers; veneer cores at the center improve screw holding power and reduce panel weight.

WAFERBOARD CORE

Sandwiching slivers of wood between hardwood face veneers eliminates voids and produces lightweight panels with reduced formaldehyde emissions.

MEDIUM DENSITY FIBERBOARD

Grind wood to dust and glue it back together with a hardwood veneer face, and you get a panel that's perfectly flat and dimensionally stable but heavy and weak.

Aircraft plywood: 1.5-mm mahogany, 4'x8' sheet, $82

Harbor Sales Co., 1401 Russel St., Baltimore MD 21230; 800-345-1712.

Flexible plywood: 9-mm Flexcurve, 4'x8' sheet, $35.50

Allied Plywood, Box 1153, Littleton, MA 10460; 508-486-8808.

Marine plywood: 18-mm Bruynzeel Regina mahogany, 4'x8' sheet, $353

M.L Condon, 260 Ferris Ave., White Plains, NY 10603; 914-946-4111.

Medium-density overlay: ¾", 4'x8' sheet, $35–50

Simpson Timber Co., Third and Franklin, Shelton, WA 98584; 800-782-9378.

Balsa-core panel: #1208, ¾" DecoLite, 4'x8' sheet, $4.26 per sq. ft.

Baltek Corp. Box 195 Northvale, NJ 07647; 201-767-1400.

Paper-core panel: ¾" Tripanel Marine, 4'x8' sheet, $110–130

Tricel Corp, 2100 Swanson Court, Gurnee, IL 60031; 800-352-3300.

Veneer-core plywood: ¾" Appleply, 4'x8' sheet, $110

States Industries, 29545 Enid Road East, Box 7037, Eugene, OR 97401; 800-626-1981.

Hybrid-core panel: ¾" Fiberply, 4'x8' sheet, $60 in bulk

Georgia Pacific, Box 1763, Norcross, GA 30091; 800-284-5347.

Waferboard plywood: ¾" Multicore, 4'x8' sheet of oak, $48, birch, $38

Weldwood of Canada Ltd., 2000 Argentia Rd., Plaza One, Suite 200, Mississauga, Ont., Canada L5N 1P7; 905-542-2700.

Fiberboard core plywood: ¾", 4'x8' sheet, natural maple, $41.60

Allied Plywood, Box 1153, Littleton, MA 10460; 508-486-8808.

DECODING A PLYWOOD LABEL

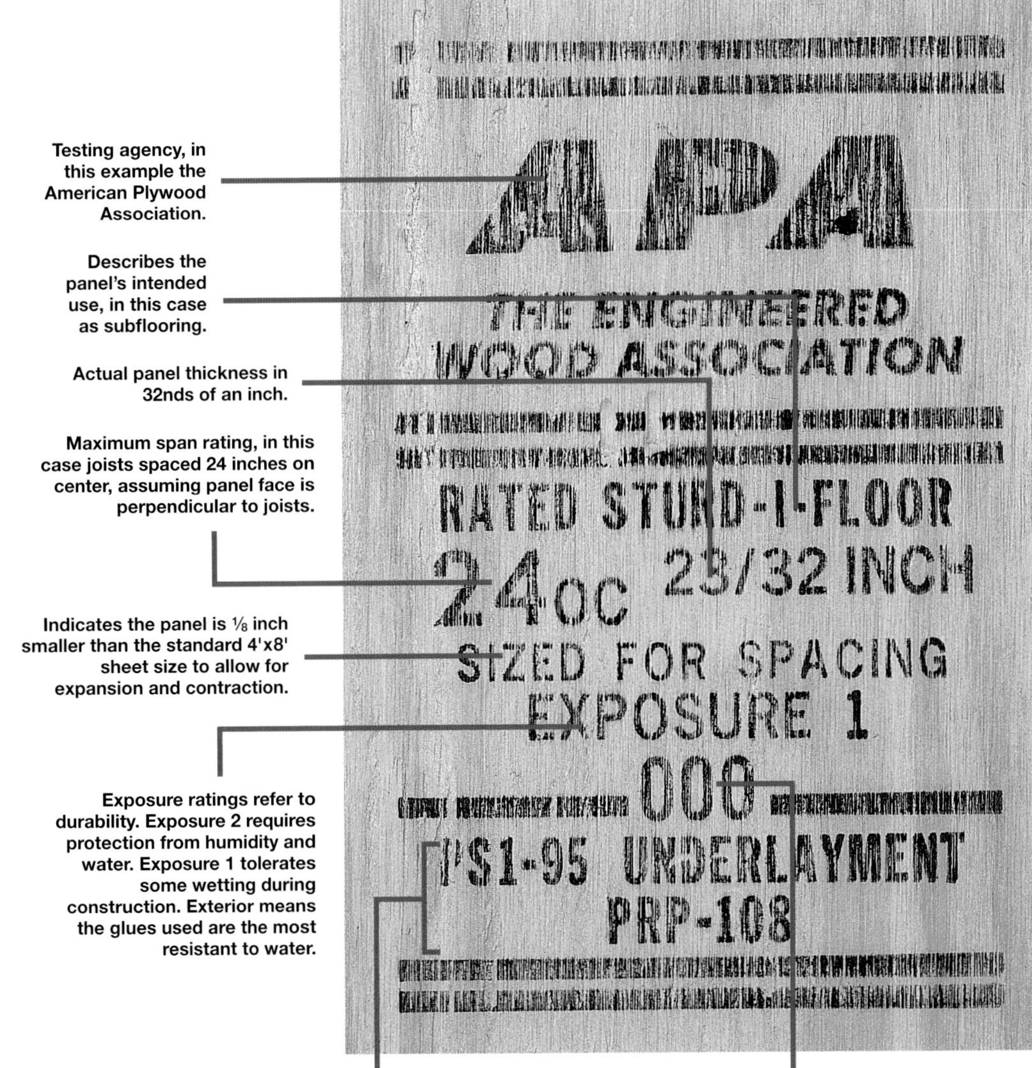

treated wood

We love wood; so do bugs and bacteria. Treated with preservatives, wood can last 30 to 40 years or even longer. Untreated, decay-prone wood can rot in a year. Builders use treated wood in foundations, garden structures, docks, swing sets—any place wood meets ground or water. In the last few years, questions have been raised about the effect of treated wood on health and the environment. The U.S. Environmental Protection Agency has issued warnings about some preservatives and gives limited approval to compounds of arsenic, used in almost all treated wood. But at a time when common garden pesticides are blamed for rising cancer rates, it is worth looking at alternatives: naturally resistant woods and man-made lumber.

REDWOOD

Redwood has natural insect and decay resistance, old-growth more so than second-growth. It is the most expensive outdoor wood in most parts of the country.

WESTERN HEMLOCK

Western hemlock, a CCA-treated species often used in western states, does not accept preservatives easily, so cuts are made in it to enable the chemicals to penetrate. Price and longevity are comparable to other CCA-treated woods.

SOUTHERN YELLOW PINE

CCA-treated southern yellow pine is the least expensive type of treated wood and by far the most widely used.

PLASTIC

Plastic lumber from reused milk containers costs 60 to 70 percent more than treated wood.

CEDAR

Cedar, also naturally insect- and decay-resistant, costs 20 to 30 percent more than treated wood.

TREX

Trex (wood-polymer composite), a blend of sawdust and recycled plastic, accepts paint or stain. It is currently the only composite lumber for outdoor use. Trex 2x4s cost $1 per foot.

NATURAL RESISTANCE

Redwood and cedar are the most common varieties of naturally rot-resistant lumber still in use, but black locust, red mulberry, Pacific yew and several others also are available. All are significantly more expensive than treated wood; however, they are less likely to check and split than treated yellow pine.

TREATED WITH WHAT?

The most widely used preservative in residential lumber is chromated copper arsenate (CCA). Ammoniacal copper arsenate (ACA) and ammoniacal copper zinc arsenate (ACZA) are applied on a smaller scale, primarily to wood species found in western states. EPA studies of health risks associated with compounds of arsenic found short-term illness following intense exposure and a few cases of acute poisoning caused by failure to use precautions in handling the wood.

Ammoniacal copper quat (ACQ) is a new product developed as an alternative for users wary of the arsenates. Considered less toxic than arsenical solutions, ACQ is subject to fewer EPA regulations; however, it has not undergone long-term testing. Borate and copper naphthenate solutions are commonly sold for home application to untreated wood, or for treating cut ends of treated wood. Both are considered safe if commonsense precautions are observed. Two older preservatives, pentachlorophenol and creosote, have been ruled too toxic for interior use. Norm Abram says he "wouldn't recommend using them at all, indoors or outdoors."

BUYING AND USING TREATED WOOD

If you plan to purchase lumber treated with CCA, ACA or ACZA, select high grades; premium wood cracks and splits less. Note the preservative retention level (stamped on the wood or a tag)—the preservative, by weight per cubic foot, that remains after treatment. A level of .40, rated for ground contact, means there is almost one-half pound of preservative per cubic foot of wood. Be wary of wood marked "treated to refusal"; the refusal zone of some species is just below the surface, leaving the inner wood unprotected.

Read the EPA's consumer information sheet (posted at the lumberyard). Its commonsense instructions: Avoid contact with food or drinking water; don't burn treated wood; avoid inhaling its sawdust; work outside; wash hands before eating or smoking.

Treated wood is shipped direct from treatment to lumberyards, so preservatives may still be leaching from the surface. When you purchase the lumber, make sure it's dry. Wear gloves when stacking or carrying treated lumber (but not when using power tools). After construction, wait two or three weeks and apply two coats of water-repellent sealer with UV inhibitors—more on cut ends. Stains and paints can be applied over sealer, but let the wood dry at least three months and try to keep it out of the sun. Prime before painting. Water-repellent sealer lasts about a year, a penetrating oil stain three to five years and a good paint job six or more.

PLASTIC LUMBER

Recycled plastics, generally high-density polyethylene from discarded milk containers, now stand in for wood in many applications. But the industry is so young that uniform test procedures and manufacturing standards are just being developed. As a result, the quality and performance of much of the plastic lumber on the market ranges from uncertain to questionable. It comes in many colors and standard lumber dimensions and can be tooled like wood. You can use power tools on it, but carbide-tipped saw blades, router bits and drills are recommended because of the material's density; we also recommend using special fasteners. Among the drawbacks is the price, more than half again as much as treated wood. Plastic lumber is heavier than dry wood of the same size.

It is not considered a structural material and should not be used in load-bearing applications. Around the house, its use is limited for now to deck and dock boards, railings, fences and landscape ties.

finials

Finials come in a wide range of materials, shapes and prices. We picked a group of 4-inch cannonballs to illustrate the point. Interestingly, there's little correlation between price and quality. Among the most expensive we found were some made of poplar, touted in a catalog for outdoor use, even though this wood has virtually no rot resistance. When we called to order the hemlock finial, we were assured it was treated to resist rot; after we saw it and rechecked, we were told it was not. Our choice from this group: the $12.50 redwood. It's a reasonable price, nicely made and less likely to split than the treated pine.

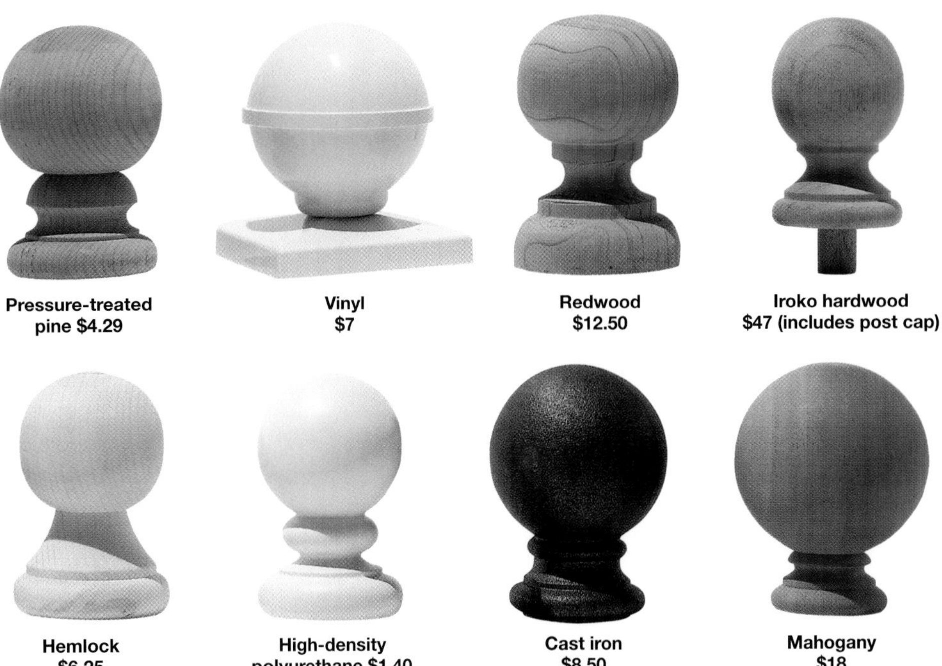

Pressure-treated pine $4.29 | **Vinyl $7** | **Redwood $12.50** | **Iroko hardwood $47 (includes post cap)**

Hemlock $6.25 | **High-density polyurethane $1.40** | **Cast iron $8.50** | **Mahogany $18**

SOURCES

Redwood
California Redwood Association, 405 Enfrente Drive, Suite 200, Novato, CA 94949; 415-382-0662.

Southern yellow pine, CCA-treated
Georgia-Pacific Corp., Box 105605, Atlanta, GA 30348; 800-284-5347.

Cedar
Western Red Cedar Lumber Association, 1100-555 Burrard St., Vancouver, BC V7X 1S7; 604-684-0266.

Trex plastic/wood composite decking
Mobile Chemical Co., Composite Products Division, 800 Connecticut Ave., Norwalk, CT 06856; 800-289-8739.

Pressure-treated pine finial
Universal Forest Products, Box 389, Windsor, CO 80550; 303-686-9651.

Vinyl finial
Heritage Vinyl Products, Hwy. 45, Box 460, Macon, MS 39341; 601-726-4223.

Redwood finial
The Wood Factory, 111 Railroad St., Navasota, TX 77868; 409-825-7233.

Iroko hardwood finial
Country Casual, 17317 Germantown Road, Germantown, MD 20874-2999; 800-284-8325.

Hemlock finial
Constantine's, 2050 Eastchester Road, Bronx, NY 10461; 800-223-8087.

High-density polyurethane finial
Georgia-Pacific, Box 1763, Norcross, GA 30091; 800-284-5347.

Cast iron finial
Lawler Foundry Corp., Box 320069, Birmingham, AL 35232; 800-624-9512.

Mahogany finial
Boston Turning Works, 42 Plympton St., Boston, MA 02118; 617-482-9085.

wood floor finishes

Jeff Hosking, who is often seen refinishing worn floors on *This Old House,* recommends always buying the most expensive finishes available. Before choosing a sheen, get a sample of gloss, semigloss or satin finishes. For a more natural wood look, use satin for the final coat. Ask for a Material Safety Data Sheet (MSDS), which lists any hazardous chemicals and necessary safety precautions. (Retailers are required by law to supply this information.) If you have any questions or concerns, call manufacturers directly to discuss products.

THE COATINGS

OIL-MODIFIED URETHANE

A combination of synthetic uralkyd resin, oils, petroleum-based thinners and metallic driers (cobalt, manganese, zirconium, zinc). The oil may be linseed (the most durable) or safflower or soybean (which don't yellow as much over time). Urethane-to-oil ratio determines softness and flexibility. Less oil results in a harder finish.

VARNISH

Natural or synthetic resins combined with oil (linseed, tung, safflower or soybean), petroleum-based solvents and metallic driers. Use only long-oil (alkyd) varnishes with 8- to 24-hour drying times. They are more flexible and longer-lasting than quick-drying short-oil (phenolic) varnishes. Products with China tung oil are easier to apply and produce a smoother finish.

PENETRATING OIL

Tung oil (pressed from the nut of the Chinese tung tree) or linseed oil (flax-seed extract) in a petroleum-distillate solvent. May contain pigments. Typically covered with a buffable paste wax.

MOISTURE-CURED URETHANE

Combines urethane resins and solvents with di-isocyanate methyl-benzene, which cures the finish by pulling moisture from the air. Also contains dangerously flammable xylene solvent and toxic ketones. Very toxic; always use a proper respirator. For professionals only.

WATER-BASED URETHANE

Resins can be acrylic (less durable), acrylic and polyurethane (very durable) or pure urethane (most durable) dispersed in water with solvents such as glycol ether, which may be extremely toxic. Cross-linkers may be added just before application to make the finish more durable. These are highly toxic and should only be used by professionals.

SHELLAC

Nontoxic flakes, from the resin secreted by lac bugs in India and Thailand, dissolved in denatured alcohol. It comes in two versions: amber (orange) or clear (white). Toxic until thoroughly dry (about 48 hours). Flakes mixed with ethanol make a nontoxic product. (Drug companies use an ethanol-based shellac to coat pills.)

Oil-modified urethane: Woodline Poly, $25 per gal.

BonaKemi USA Inc., 14805 E. Moncrieff Pl., Aurora, CO 80011-1207; 800-872-5515.

Varnish: BenWood Satin Finish, #404, $23.78 per gal.

Benjamin Moore & Co., 51 Chestnut Ridge Rd., Montvale, NJ 07645; 800-344-0400.

Water-based urethane: Pacific Ultra, $90 per gal.

BonaKemi USA Inc.

Shellac: Bulls Eye Amber #701, $18.50 per gal.

Wm. Zinsser & Co., 173 Belmont Dr., Somerset, NJ 08875-1285; 908-469-4367.

Prefinished flooring, ¾" thick, 3¼" face width, oak with semigloss finish; $6.60 per ft.

Mirage, 1255 98th St., Saint-Georges, Beauce, Quebec, CN G5Y 5C2; 418-227-1181.

FOR MORE INFORMATION

Hardwood Floors, 6 issues, $36

National Wood Flooring Association, 1846 Hoffman St., Madison, WI 53704; 608-249-0186.

ADDITIONAL SOURCES

Flocked applicator: Floor Coater, #6118, 18", $25.75

Padco Inc., 2220 Elm St. SE, Minneapolis, MN 55414; 800-328-5513.

Paste wax: Butcher's Boston Polish, 1-lb. can, $6

The Butcher Co., 67 Forest St., Marlborough, MA 01752; 508-481-5700.

PREPARING THE WOOD

To finish raw wood or replace an existing finish, you must prepare a perfect surface. First sand the floor with 36-grit paper, then 80-grit, making one pass with each using a drum sander and moving with the grain. Where the drum can't reach, sand with an edger. Sand only until the grain is smooth and clear of old residue. Scrape corners by hand. Vacuum after each sanding. (Sand parquet floors at a 45-degree angle to the squares, with the final pass parallel to the squares.) Finish with a 100-grit screen on a floor-buffing machine.

To recoat an existing oil-based finish, wash the floor first with turpentine to remove any wax, grease or dirt. (For water-based urethanes, use a cleaner recommended by the manufacturer.) Then screen with a 120-grit abrasive on a buffer, vacuum thoroughly and wipe with a tack rag. Apply one or two coats of the same brand and type of finish. On shellacked floors, remove wax with turpentine, apply a fresh coat or two of shellac and protect with wax.

FINISHING TOUCHES

Oil-modified urethane is Jeff Hosking's all-around favorite. Leaves a thick, honey-toned surface that darkens with age. It is highly scratch-resistant once cured (up to four weeks after application), withstands water and is relatively easy to apply. Hard to touch up and recoat; incompatible with most other finishes. Jeff applies three coats using a 6-inch bristle brush or a pad applicator.

Varnish is one of the most beautiful natural finishes, although it lacks the durability of a urethane. Oils give it a rich amber finish that yellows with time. Hosking rarely recommends varnish but says it goes on fairly easily with a 6-inch natural-bristle brush. (Badger hair is best.) He always puts down three coats, then waxes.

Wood floors that really look like wood are finished with a **penetrating oil** that is relatively easy to apply with a natural-bristle brush. It is time-consuming to maintain because its wax protection is easily worn away. Comes in many colors and shades. Spilling foods or common household products containing alcohol, vinegar or ammonia can create disastrous discoloring.

The hardest, longest lasting, least yellowing (if aliphatic solvents are used instead of aromatic), most stain-resistant finish on the market, **moisture-cured urethane** is extremely difficult to apply, deadly to breathe and so volatile that a single spark can cause an explosion. Floors may look as though they are encased in plastic. Absolutely not for do-it-yourselfers.

The most popular floor finish, **water-based urethane** dries fast—in as little as an hour—making application convenient but tricky. Little or no odor or yellowing. To keep it from raising the grain on raw wood, Hosking uses a sealer coat of thinned shellac, then applies three coats of finish with an 18-inch roller applicator. Contains toxic chemicals; always wear a respirator.

Shellac gives a rich, warm appearance to wood. Its quick drying time can leave lap marks. Hosking suggests putting down three coats, then waxing. When thinned (a gallon of alcohol to one of shellac), it makes an excellent sealer under other finishes. Test compatibility first. Shellac is susceptible to water spots but easy to touch up. It is the least toxic finish and moderately durable.

All samples are white oak. Jeff Hosking applied three coats of finish to each, except the bottom one, which got one coat of oil and one of wax.

PREFINISHED FLOORING

"With prefinished floors," says Hosking, "there's no sanding, no awful chemicals to breathe, no bother." But he doesn't like to install them "unless it's in an area of very light traffic." In his experience, most factory finishes don't hold up well. The rounded or beveled edges on prefinished planks leave dirt-trapping grooves, something rarely found on a jobsite sand and finish. Hosking is even less complimentary about prefinished veneer flooring: "Veneer is only acceptable in a room you hardly use. You sand it once and you're very lucky if you don't see plywood come through."

notes

roofing, siding and windows

CLAY ROOF TILES | GUTTERS

CEDAR SIDING | WINDOW GLASS | LIGHT PIPES

CORROSION-RESISTANT SCREENS

clay roof tiles

Made from pulverized clay mixed with water, clay roof tile is shaped by pressing the mixture into a mold or feeding it through an extruder, which produces a ribbon of molded clay that's then cut into pieces. The pieces are fired in a kiln that can reach temperatures of 2,400 degrees Fahrenheit. The higher the kiln temperature, the harder and more durable the tile. Soft, porous tiles are fine for hot, dry regions but would disintegrate in northern freeze-thaw climates.

Tiles are either curved (pantiles) or flat, resembling shingles. Different clays yield different colors, depending on the presence of certain elements. Iron oxide, for example, produces the characteristic terra-cotta red; manganese produces black or dark purple. Glazes may be sprayed over tiles to create a glossy effect.

THE COST OF CLAY-TILE ROOFING

New clay tile starts at about $300 per square (a square is a 10-by-10-foot area), labor not included. That's slightly less than copper ($400 per square) and about the same as slate. Cedar shingles and steel roofing cost about $150 per square. Composition shingles are the most affordable, about $30 per square for a 25-year roof. At around $800 per square, salvaged tiles are expensive but still less costly than having new tiles custom made to match an old design. Manufacturers charge setup fees for one-of-a-kind tiles that range from $300 to $3,000, depending on intricacy. Labor costs vary according to the pitch of the roof and the size of the job, but laying tile, slate and cedar is slow, painstaking and pricey. Installing copper or metal costs about a third as much. Composition shingles can be installed by homeowners. Life expectancy for each type of roofing varies according to climate, the shape of the roof and the quality of the materials. On average, warranties cover cedar for 20 years, composition shingles for up to 30 years, steel for 30 years and slate, tile and copper for 50 years.

SOURCES

Unglazed Gothic tile
Renaissance Roofing, Box 5024, Rockford, IL 61125; 800-699-5695.

Natural-clay red flat tile
Renaissance Roofing.

Spanish tile with wavy profile
The Tile Man, Route 6, Box 494-C, Louisburg, NC 27549; 919-853-6923.

Fluted French tile
The Tile Man.

Spanish tile with Brookville green glaze
Renaissance Roofing.

Conosera unglazed tile
The Tile Man.

French lavender glass flat tile
The Tile Man.

French Gothic fluted tile with Brookville green glaze
Renaissance Roofing.

Colonial textured tile
The Roof Tile & Slate Co., 1209 Carroll, Carrollton, TX 75006; 800-446-0220.

Dresden tile
The Tile Man.

FURTHER READING

"The Preservation and Repair of Historic Clay Tile Roofs," by A.E. Grimmer and P.K. Williams, Preservation Briefs No. 30
U.S. Dept. of Interior, Nat'l Park Service, Box 37127, Washington, DC 20013; 202-208-7394.

SALVAGED TILES FOR OLD OR NEW ROOFS

1. 1890s unglazed Gothic tile, natural-clay red, $1,250 per 10-by-10-foot square.
2. 1890s natural-clay red flat tile, $1,100 per square.
3. Early 1900s Spanish tile with wavy profile, $225 per square.
4. Early 1900s fluted French tile, highly glazed fox red, $500 per square.
5. 1920s Spanish tile with Brookville green glaze and antimony spots, $650 per square.
6. Late 1890s Conosera unglazed tile, $350 per square.
7. 1890s French lavender glass flat tile, used for skylights, $100 per tile.
8. Early 1900s French Gothic fluted tile with Brookville green glaze, $850 per square.
9. 1920s Colonial textured flat tile (coated with lichen), $550 per square.
10. 1930s Dresden tile with wire-scored or combed surface, $350 per square.

gutters

It's hard to get excited about gutters. They seem like one of those elements that just comes with the house, like walls and rafters and the front steps. But live through a rainy season without them and you realize how helpful, if not exciting, gutters are. Moisture is always out to attack your house. Water running off the roof ruins the paint, warps and rots the doors and window frames and erodes the soil around the foundation. Gutters capture water and direct it to downspouts, which conduct it safely away from your home. In regions where rainfall is light and porous soil drains water quickly, you can get by without gutters. But in much of the country, a house without gutters is asking for moisture problems.

INSTALLATION

Gutters aren't standard on all homes. Some houses are designed with deep eaves that broadcast rainfall away from the house. Good grading and drain tile (perforated pipe) buried around the perimeter of the house also disperse water. If these precautions aren't taken, problems are inevitable. The house pictured here, built without gutters, is three years old and already the wood window sash and garage doors are rotted. The builder replaced the decayed wood, and subcontractors installed copper gutters.

1. Nothing looks worse than a gutter that's askew. "You can see it from a block away," says installer Stephen Lancaster. The solution: Plan the installation, measure carefully and snap a chalk line along the fascia as a guide. Some slope—1 inch per 20 feet of run—is necessary so water flows to downspouts. Seams and downspouts should be symmetrical and unobtrusive. **2.** Copper makes great gutters. It's handsome, it won't rust and, with proper care, will last as long as the house. Because this roof is steep and large, Lancaster used a 6-inch-wide trough instead of the more typical 5-inch. The back of the gutter is taller than the front so water will spill away from the house. **3.** Copper is trickier to install (and therefore more expensive) than other materials. To assemble a corner, a multistep process, one end is cut to size with tin snips. Tabs are cut into the other end and bent into place.

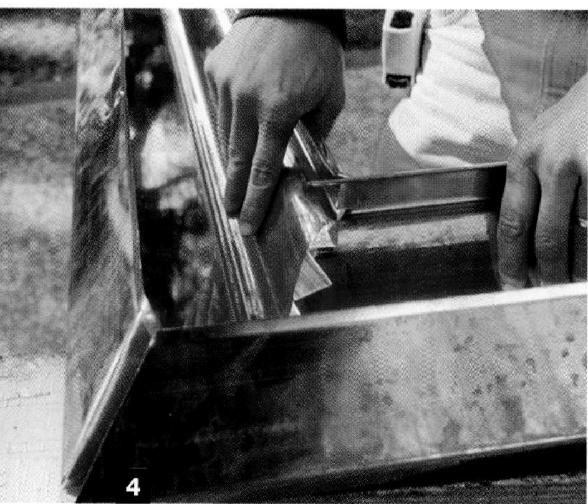

4. After fitting the corner pieces together, joints are riveted and soldered. **5.** Downspouts are connected to the troughs with drop tubes. These are soldered to keep them firmly in place and to prevent water from leaking through the joint. **6.** Sections of gutter are assembled on the ground and lifted into place. This saves the installer from hanging in the air to solder joints.

Aluminum gutters: $5–9 per lineal ft. installed
Alcoa Building Products, Box 3900, Peoria, IL 61612; 800-962-6973.

Copper: $15–20 per lineal ft. installed
South Side Roofing & Sheet Metal, 290 Hanley Industrial Ct., St. Louis, MO 63144; 314-968-4800.

Galvanized: $6–10 per lineal ft. installed
South Side Roofing & Sheet Metal Co.

Wood: $12–20 per lineal ft. installed
Blue Ox Millworks, 1 X St., Eureka, CA 95501-0847; 800-248-4259 or 707-444-3437.

Plastic PVC Snap-Seal Contemporary: $3 per 10 ft.; Snap-Seal Traditional, $5 per 10 ft.
GSW Thermoplastics Co., 1735 Highwood E., Building J, Pontiac, MI 48340; 800-662-4479.

LeafGuard: $7 per lineal ft. installed
Englert Inc., 1200 Amboy Ave., Perth Amboy, NJ 08862; 908-826-8614.

Rainhandler: 5-ft. with mounting materials, $21.90
Savetime Corp., 2710 North Ave., Bridgeport, CT 06604; 800-942-3004 or 203-382-2991.

Gutter Helmet: $9 per lineal ft. installed
American Metal Products, Masco, 8601 Hacks Cross Rd., Olive Branch, MS 38654; 800-423-4270.

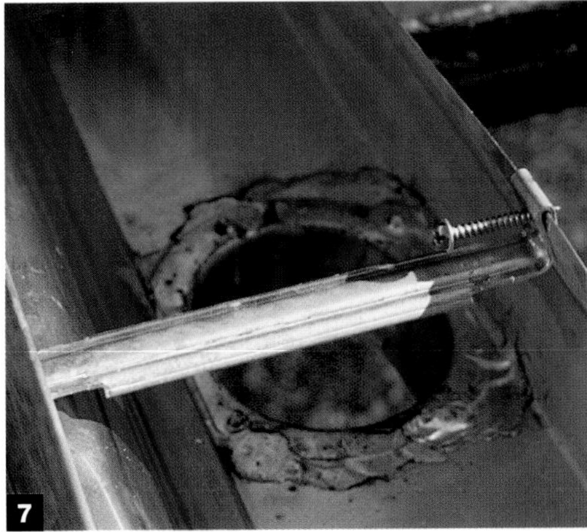

7. Fasteners are designed to hold the gutter securely to the fascia or the eaves, even when the troughs are loaded with snow and debris. They come in a range of configurations; the best one for the job depends on what type of gutter is used and what the drip edge and the roofline are like. Lancaster selected fascia-mounted hangers because they don't disturb the roofing material. Unlike straps, which wrap around the exterior of the gutter, these brass hangers are invisible once installed. They're slid into place before the gutter is mounted. **8.** The gutter is attached to the fascia just below the drip edge at 36-inch intervals. Lancaster uses 1¼-inch stainless-steel screws.

9. The gutter above this bay window includes several odd angles and awkward jogs. Measuring, soldering and installing this small section took two workers half a day to complete. The straight runs go more quickly—an average-size house can be completed in a day or less. **10.** The size and frequency of the downspouts is determined by the area and steepness of the roof. As a rule, for every 100 square feet of roof area, add 1 inch to the diameter of the downspout. A 3-inch-diameter downspout can carry rainfall from 300 square feet of roof, for example. Downspouts should be spaced a maximum of 40 feet apart to prevent water from backing up.

11. Soldering the seams on these elbows prevents water, which always takes the path of least resistance, from leaking out and invading the eaves. Aluminum, plastic and wood gutters rely on caulk to keep joints watertight. **12.** Screening keeps leaves, pine needles and granules of asphalt roofing from collecting inside the gutters.

ALUMINUM

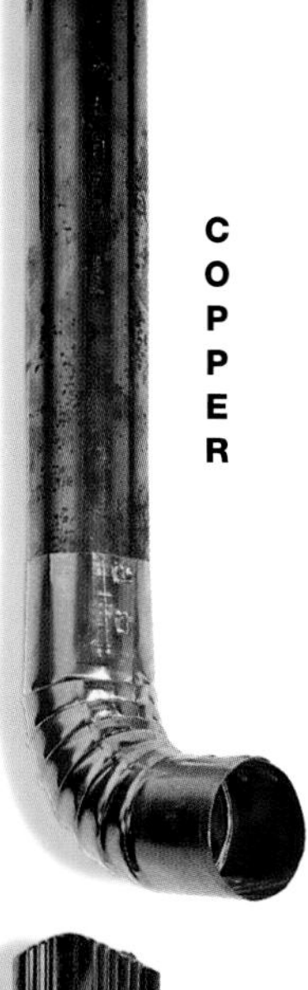

COPPER

GALVANIZED

GUTTER CHOICES

Selecting gutters is just slightly less complicated than choosing wallpaper. You'll be amazed at the range of materials, prices, profiles, colors and fasteners. Aluminum gutters are generally available in three sizes, four stock profiles and a dozen colors, including a new finish from Alcoa Building Products that imitates copper but costs only $9 per lineal foot, installed (versus $15 to $20 for the real thing). Here's an overview of common gutter and downspout materials. Prices are suggested retail, include installation and are based on the lineal foot.

ALUMINUM

$5 to $9. Select the thicker variety (.032 inch) for greater stability and strength.

WOOD

$12 to $20. Redwood or Western red cedar will last 50 years or longer if oiled regularly.

PLASTIC

$2 (installed by homeowner) to $7.50. Low-cost but not very sturdy; can crack in cold weather and deform in heat.

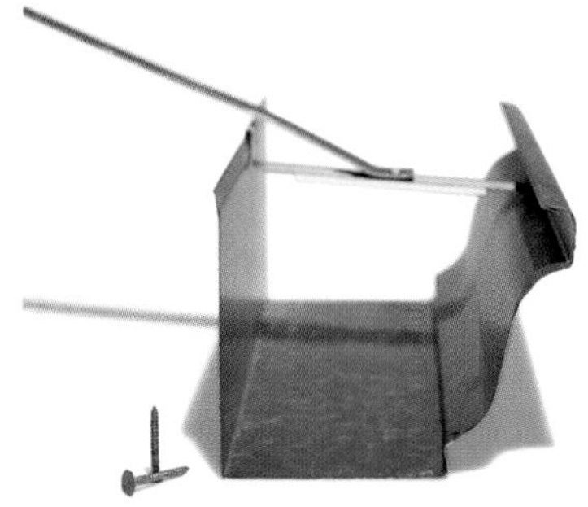

GALVANIZED

$6 to $10. Steel coated with zinc (the galvanizing layer), these are sturdy but prone to rust.

KEEPING THE MUCK OUT

Cleaning gutters is loathsome. What could be less fun than dipping your hands into a sodden pile of dirt, decomposed leaves and who knows what else while swaying on a ladder? Manufacturers have caught on to the fact that many people will pay dearly for a product that keeps them out of their gutters. LeafGuard and Gutter Helmet take advantage of surface tension to pull water into the gutter while deflecting debris. Both are made of aluminum. Gutter Helmet is installed over existing gutters, while LeafGuard is a complete gutter system. Rainhandler isn't a gutter—it's a "rain dispersal system." That means the aluminum louvers disperse water as it falls from the roof, projecting it away from the house. Leaves simply wash away. Do any of these products actually work? Norm Abram is skeptical. "They have yet to invent a maintenance-free gutter," says the *This Old House* master carpenter. "When they do, I'll buy it."

LEAFGUARD

$7 per lineal foot, including installation.

RAINHANDLER

$4.50 per lineal foot, not including installation.

GUTTER HELMET

$9 per lineal foot, including installation.

Left: Clogged gutters give new meaning to the term "rooftop garden." Right: Painting galvanized gutters helps prevent rust. To make paint stick, scrub the gutters with scouring pads and mineral spirits or a liquid deglosser to remove oils and give the surface tooth. Wipe with a rag and let dry for a couple of hours. Then apply a primer for galvanized steel and let stand overnight. Solvent-based primers work better as a rule. Finally, test the coating with a fingernail; if it doesn't scratch, it's ready for paint.

MINDING THE GUTTERS

Gutters do more harm than good if they're leaky or jammed. If it can't flow out, water backs up under the roof and makes its way into the house. Clear your gutters every spring and fall—take your garden hose up with you and flush the troughs clean or use a child's rake, a rubber spatula or your fingers. Use the hose to flush out the downspouts. If they're clogged, clear them with a plumber's snake. Inspect the gutters for holes and cracks; if you find any, call the manufacturer for advice on repairs. (Leaks at joints and seams can be filled with caulk.) Wood gutters should be coated with wood preservative or thinned raw linseed oil to prevent cracking. Sagging gutters are the result of broken fasteners. You can buy new ones at a home center, but be sure they're compatible with your gutter (some types of metals react). If you find yourself skipping or postponing maintenance, hire someone to do it. Many gardening and remodeling companies clean and repair gutters. They'll come to your house twice a year, do the work and bill you $100 or so, depending on how high the gutters are, how big your house is and what kinds of repairs are needed. It's worth it.

cedar siding

Clapboard is the quintessential American weather shield. All across the land, long thin strips of overlapping, tapered wood planks cast crisp horizontal shadow lines on Victorians, Cape Cods, farmhouses and Colonials. The oldest frame building in North America—the circa 1637 Fairbanks house in Dedham, Massachusetts—wears a clapboard skin. Newer houses in developments also have the look of clapboard, although the siding often turns out to be a veneer of vinyl or aluminum.

Clapboard goes on fast, presents a solid surface to the weather, doesn't trap moisture and is easily repaired. It's expensive—twice as costly as vinyl—but high-quality clapboard can last for centuries.

Like many features of a well-built house, clapboard appears simple to install. But to do it properly, every joint and edge requires multiple layers of protection to keep out water, and to look right, clapboards must be parallel, consistently spaced and aligned at the corners. *This Old House* contractor Tom Silva has been siding with clapboard for three decades. "I love the look of it," he says. "There's nothing nicer than working with good materials."

Before he can begin nailing up siding, Tom must prepare the surface beneath. "If the sheathing moves while I'm nailing, the clapboards will crack." He reattaches loosened sheathing to the studs and replaces any rotten wood. Then he marks the location of each stud on the water table (the horizontal trim board at the bottom of the wall). The marks will show him where to nail through the clapboards.

Because wind-driven rain can work its way between and behind lapped siding, Tom always staples wide sheets of housewrap over the sheathing. Then he slips 6-inch-wide strips of 15-pound roofing felt behind all the vertical trim pieces at corners, windows and doors and caulks between the felt and trim with an acrylic latex sealant. All horizontal trim is covered with 4-inch-wide strips of lead flashing to keep out water draining down the side of the house. The flashing is nailed to the sheathing with galvanized roofing nails. Tom overlaps all flashing joints 2 inches or more and seals them with caulk.

To begin the siding work, he nails a 1-inch-wide wood strip across the bottom of the wall. Cut from the narrow edge of a clapboard, the strip cants the first piece of siding to the same angle as the boards that follow. With a few taps of his hammer, he sinks 5d stainless nails about 3/8 inch above the thick bottom edge of the first siding board and into the studs.

CLAPBOARD INSTALLATION

PREPPING

When re-siding an old house, Tom Silva first taps his hammer over the sheathing, listening for dull thuds that indicate rotten wood and loose boards. Solid thumps locate studs. After all the punky boards are replaced, he shoots a few extra nails through the sheathing into the studs.

WRAPPING

Spun olefin housewrap over sheathing stops water and wind. It also lets interior moisture escape, "just like a Goretex jacket," Tom says. When cutting wrap around a window he leaves enough extra to tuck in behind the trim.

WEATHERPROOFING

If water gets behind clapboard, rot will run riot through sills and sheathing. Left: With three barriers in place—housewrap, felt and flashing—Tom pumps a fourth, a bead of caulk along the corner board. Right: A 1-inch strip nailed to the flashing gives the first clapboard the same cant as the boards that follow. Tom makes his own strips by ripping down the thin edge of a clapboard with a table saw.

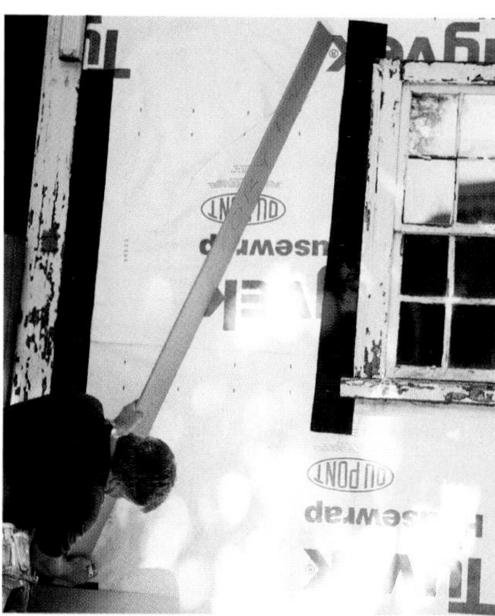

MARKING

Rather than fiddle with a tape measure, Tom uses a homemade story pole to figure overlap. The marks he transfers from the pole to the housewrap show him exactly where to place the top of each board.

FITTING

To shape the end of a piece of clapboard to a windowsill corner, Tom first holds the board against the side of the sill to mark its height. Then he places the board on top of the sill to mark how far it extends beyond the edge of the window. Finally, he cuts out the notch defined by the intersecting lines.

MOVING UP

A clapboard wall swiftly acquires its characteristic look as Tom nails course after course to the sheathing. He uses lead flashing at the head casing on the window. "If you only rely on caulk to keep water out, you're asking for trouble," he warns.

SOURCES

Factory-primed cedar siding: 6" clear vertical-grain, 66 cents per lineal ft.
MacMillan-Bloedel, 925 West Georgia St., Vancouver, BC V6C 3L2; 604-582-2690

1¾" 5d stainless-steel siding nails: $99 for 25 lbs.
Prudential Building Materials, 71 Milton Street, East Dedham, MA 02026; 617-329-3232

Acrylic latex caulk with silicone: Alex Plus, $2 per cartridge
DAP, Box 277, Dayton, OH 45401; 800-327-3339

Olefin housewrap: Tyvek HomeWrap, $165 per 9'x195' roll
DuPont Co., 107 Market Street, Wilmington, DE 19898; 302-774-1000

Cement-based clapboards: Hardiplank
James Hardie Building Products, 10901 Elm Avenue, Fontana, CA 92337; 800-942-7343

FOR MORE INFORMATION

Western Red Cedar Lumber Association
1100-555 Burrard St., Vancouver, BC V7X 1S7; 604-684-0266.

Western Wood Products Association
Yeon Building, 522 SW Fifth Ave., Portland, OR 97204; 503-224-3930.

Tempting as it is to continue, Tom stops now to plan a layout. He wants all the boards parallel and dead level, with a consistent overlap. And he wants the bottom edge of the boards at the top and bottom of the window to line up with the window trim. While some carpenters measure the space and laboriously figure and refigure the number of boards that fit, Tom's way is simpler. He makes a story pole. Measuring from one end—the "bottom" of the pole—he marks up every 4½ inches, the most a 6-inch-wide clapboard should be exposed.

To use the story pole, he snaps a horizontal chalk line one clapboard's width up from the bottom of the windowsill. Next, he rests the pole's bottom on the first clapboard and, holding the marked edge against the wall, angles the pole until one of the marks touches the chalk line. Then he transfers every mark below the chalk line from pole to housewrap. Both sides of the window are story-poled so he can snap chalk lines between each pair of marks. All he has to do is hold the top of each clapboard to the line and nail away.

Ideally, each piece of clapboard should be long enough to reach from one side of a wall to the other without a seam. When it can't, Tom makes an overlapping scarf joint and beds it in caulk. Just the top board on the scarf gets nailed near the joint. "It's the only time I nail into the sheathing without hitting a stud," he says. He makes sure joints never line up one above the other and that they are spaced randomly.

Once he has fit the clapboard to the windowsill, Tom picks up the story pole again to determine the spacing to the top of the window. Then he snaps more lines and keeps nailing away, following the same sequence until he runs out of wall. With the clapboards ready for painting, Tom steps back and looks at the job. "That'll last a good long time," he says.

SIDING CHOICES

Cedar siding comes in two basic varieties: bevel siding, in which boards are sawn to create a tapered profile, and pattern siding, where the face of the board is milled into a desired shape or pattern. With bevel siding, weather-tightness comes from each board being overlapped by the one above. On pattern siding, the edges of the boards are milled so they interlock with either tongue-and-groove or lap joints. Both types of siding come with a smooth (surfaced) face and a rough (sawn) face. Bevel and pattern sidings are graded at the mill based on the defects and grain patterns in the wood. Grading for bevel siding is commonly done on the smooth side, so expect more defects if you plan to install it rough side out.

Bevel siding
Clear, vertical-grain heart

Tongue-and-groove siding
Clear heart

Wavy-edge bevel siding
Select knotty

Channel (lap) siding
Proprietary grade (knotty)

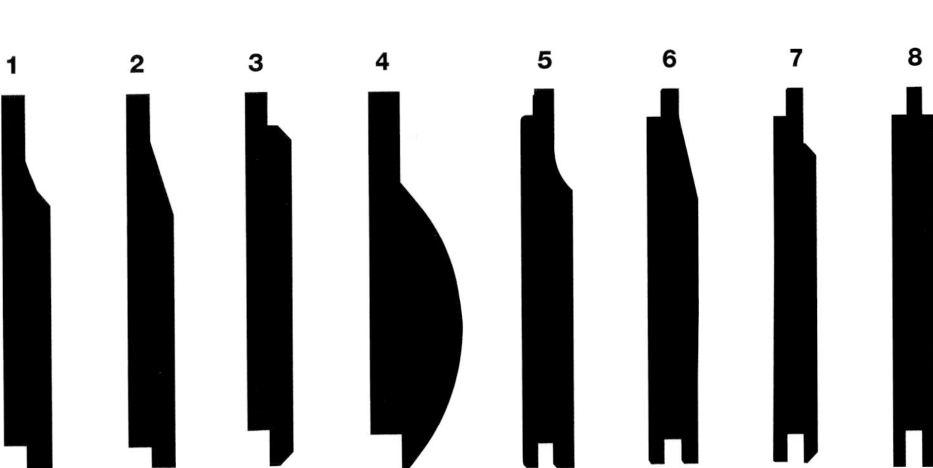

Pattern siding (with milled faces) is either lap or tongue-and-groove. Lap sidings (1 through 4) are milled to overlap the adjacent piece. For tongue-and-groove sidings (5 through 8), the pieces interlock. Patterns are identified by standard names or numbers.
1. No. 105 **2.** No. 101
3. WP-11 **4.** log cabin
5. No. 106 **6.** No. 102
7. WP-6 **8.** center-matched tongue-and-groove.

Siding: Tom Silva uses only the best—clear, all-heart, vertical-grain boards cut from 150- and 200-year-old Western red cedars. This grade costs about 12 percent more than the next best grade, called A clear, but it looks better, holds paint better and is unlikely to warp or cup. To calculate how many square feet he needs, he measures the total wall area (length times height), multiplies by 120 percent and subtracts the square footage of doors and windows.

Painting: Tom prefers factory-primed clapboards. Priming protects all sides of the boards from moisture, including the often forgotten back. Moisture passing from the house to the exterior through unprimed siding tends to lift and bubble paint and cup boards. Tom instructs painters to brush on an additional coat of oil-based primer (to seal nail holes and joints), fill the nail holes with caulk and finish with two topcoats of acrylic latex paint.

Nails: Tom says stainless steel is the only way to go. Galvanized nails leave black streaks in cedar and rust long before the boards need replacing. Blunt-tipped, ring-shank siding nails reduce splitting and provide the best holding power.

Caulk: Use lots of it around joints and trim, Tom says. Go for a siliconized acrylic latex caulk with a long warranty.

Housewrap: Spun olefin is best, in Tom's view, for keeping out wind and rain. The fewer seams the better, so he buys it in 9-foot-wide rolls and wraps it around corners and over windows. At seams, he overlaps the sheets by a couple of feet and seals them with plastic tape. Any wrinkles remaining are smoothed, folded over and stapled flat.

Flashing: Sheet lead is Tom's favorite because it's so easy to work. "I can bend it without a metal brake," he says. Copper and painted aluminum flashing are other long-lived alternatives.

window glass

Today, homeowners have a lot to choose from when they install new windows or retrofit old ones. Those who go with standard glass, whether or not they realize it, get a product with a slight greenish cast (because of its iron content). The glass will let in most of the sun's energy: 90 percent of visible light, 80 percent of ultraviolet (the short-wave radiation that degrades fabrics) and 79 percent of infrared (longer-wave radiation that causes objects to heat up). It will break into shards if struck by a baseball or a burglar or subjected to the intense heat of a fire.

Other glasses are far different. Low-emissivity glass, developed in the 1980s, offers the most dramatic temperature control with the least change in appearance. Made by coating glass with an invisibly thin layer of silver or tin oxide, this type allows most visible light to pass through but reflects interior heat, keeping rooms warmer in winter. A variety of "low-e" glasses are available for different situations. Where summer cooling costs are most significant, buyers can choose glass that deflects most of the sun's heat. But where free solar heat is an asset, they can use glass that admits most of it. Windows with different characteristics can be chosen for different exposures of a house.

Low-e coatings can be either pyrolytic, which means the metals are applied while the glass is still molten, or added later. The latter are often called "soft" coatings because they scratch more easily. Since soft coatings tend to give better thermal performance, they're good inside double-pane insulated windows. Pyrolytic coatings are the only way to go for storm or single-glazed windows.

Other, older ways of limiting heat from the sun—tinted and reflective glass—are not as well suited to homes because they change a window's look. But styles may change. PPG is conducting a marketing survey to find out what range of tinted glass homeowners might accept if they knew the benefits. Glass with slightly more green than normal, for example, admits most visible light (83 percent) but blocks nearly half of ultraviolet and 45 percent of infrared.

Besides solar considerations, glass can be selected to resist breakage, muffle sound or withstand fire. Often this is done by laminating thin plastic film between two sheets of glass. By using low-e, tinted or reflective glass for one or both of the layers, it's possible to get glass that has several advantages. This is true with insulated glass too, which has an air layer between the panes to keep heat from moving by conduction (physical contact) and convection (air flow). The right insulated glass can deal with radiant heat and security issues as well. Not all companies offer all options, so shop around.

SOURCES

Standard window glass: $1.50 per sq. ft.
PPG Industries, 1 PPG Place, Pittsburgh, PA 15272; 800-377-5267 or 412-434-5267.

Starphire low-iron glass
PPG Industries.

Nonreflective glass: Amiran, 4 mm., $25 per sq. ft.
Schott Corp., 3 Odell Plaza, Yonkers, NY 10701; 914-968-8900.

Standard heat-reflecting glass: Solarcool Bronze, $8 per sq. ft.
PPG Industries.

Low-emissivity heat-absorbing glass: bronze color, $20 per sq. ft.
Rosen-Paramount Glass Co., 45 E. 20th St., New York, NY 10003; 800-287-6736.

Insulated glass: ½", $10 per sq. ft.
Rosen-Paramount.

Heat-Mirror glass: Insulated, $8-15 per sq. ft.
Southwall Technologies, 1029 Corporation Way, Palo Alto, CA 94303; 800-365-8794.

Laminated glass: California Series laminated glass, $10-17 per sq. ft.
Southwall Technologies.

Solex (green) and Solargrey, $7 per sq. ft.
PPG Industries.

Lexan scratch-resistant plastic: MR 5000, ⅛", $5.50 per sq. ft.

Lexan Thermoclear with insulating air channels, 8 mm., $2.25 per sq. ft.
Commercial Plastics, 98-31 Jamaica Ave., Richmond Hill, NY 11418; 718-849-8100.

Privacy Glass: $80–90 per sq. ft.
Marvin Windows & Doors, Box 100, Warroad, MN 56763; 800-346-5128.

SuperLite I, $20-40 per sq. ft., and SuperLite II, $45 per sq. ft.
SAFTI, O'Keeffe's Inc., 75 Williams Ave., San Francisco, CA 94124-2693; 415-822-4222.

WINDOW GLASS OPTIONS

"CLEAR" GLASS

Standard window glass (A) is what we're used to. It's cheap and in warm climates is still the best choice for windows that don't get a lot of sun. We're so accustomed to it that we barely notice its greenish tint. But for use in solar collectors, **low-iron glass (B)** is better. It has no green, and more solar energy gets through. If you don't want to see yourself when you peer through a window, pick glass with a **nonreflective coating (C).**

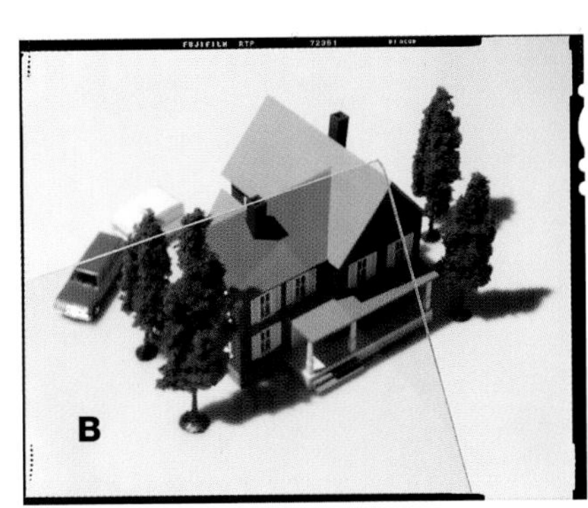

REFLECTIVE GLASS

Standard reflective glass (A) can keep sun-drenched rooms from roasting. But it darkens interiors and makes windows look like mirrors. **Low-emissivity glass (B)** is great for homes. It lets in almost as much light as standard glass but reflects interior infrared energy to keep rooms warmer in winter. It can be made to deflect or admit infrared from outside. The only drawback: It's more reflective than we're used to.

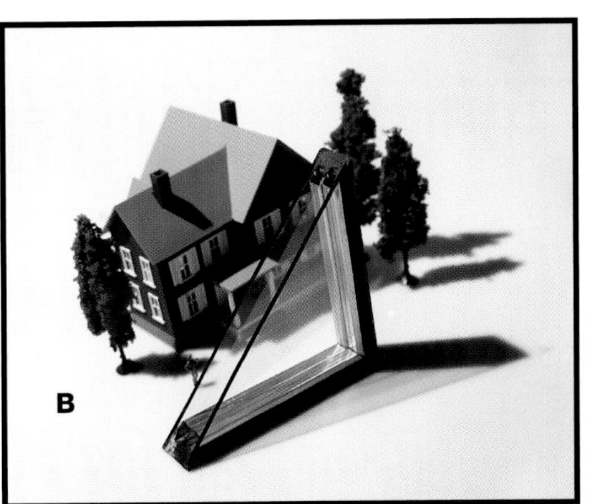

INSULATED GLASS

Insulated windows (A), which can be made of any kind of glass, help keep heat in (or out) of a building. Metal spacers filled with desiccant hold the two panes apart, creating an air layer in the middle. That space, sometimes filled with argon or other inert gases instead of just air, acts like insulation in a wall. **Heat-Mirror glass (B)** insulates twice as well because it's made by suspending a transparent sheet of plastic (with low-e coating) in the air space, creating two insulating layers.

SAFETY GLAZING

Where people might fall through, building codes require glazing that won't cause injuries if it breaks. Tempered glass is one option. Another is **laminated glass (A),** with a thin layer of polyvinyl butyral between two layers of glass. This sample has a low-e film between two PVB layers. **Lexan (B),** a plastic, is a low-cost alternative. Buy it only with a scratch-resistant coating. It's available with **insulating air channels (C),** but they're very visible.

SWITCHABLE GLASS

This is the perfect glass for folks who want a conversation piece or hate closing drapes. Switch it on and **Privacy Glass** changes from opaque to clear **(A).** Turn it off and it's back to milky white **(B).** Its magic is a thin film of liquid crystals. Usually in random order, they line up when the film is charged by electricity, letting you see through the window. The glass has no significant energy savings. Could glass be made to change from admitting to blocking the sun's heat at the flip of a switch? Maybe, but don't look for it soon.

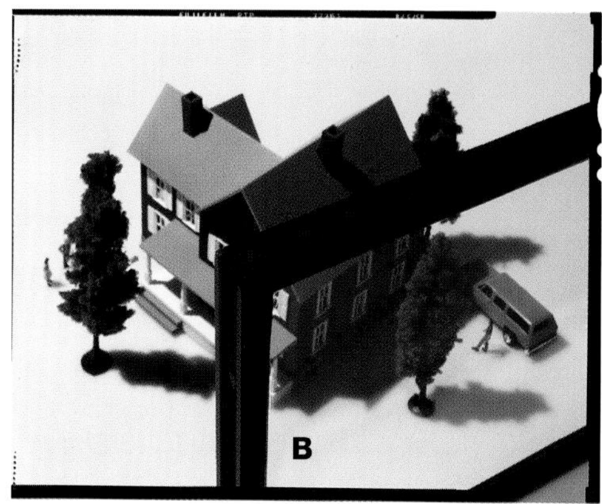

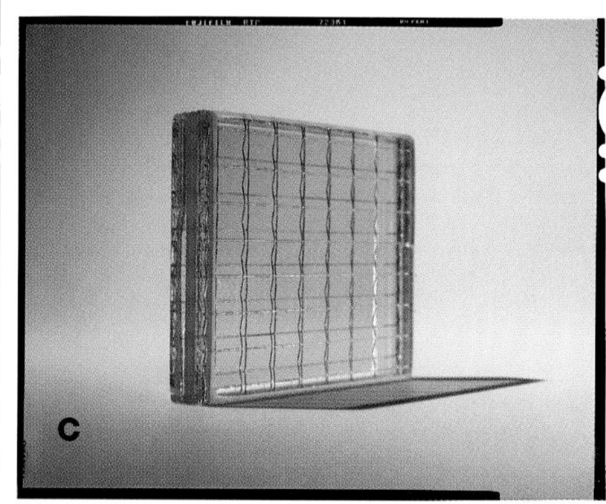

HEAT-RESISTANT GLASS

A regular window can burst during a fire, allowing flames to spread. In institutions and other places where it is crucial for glass to hold together, a traditional solution is wire-embedded glass. But wire glass and alternatives like wire-free **Superlite I (A)** can still transmit enough radiant heat as flames encroach to set a room on fire from the inside. Not **Superlite II (B).** Its inner gel layer, made of a polymer containing salt and water, actually absorbs heat. After 60 or 90 minutes, depending on thickness, the nonfire side will remain below 250 degrees. Perhaps worth buying for homes in fire-prone areas, but other fireproofing is more important. **Inferno-Lite (C)** is the ultimate in wire glass. Laminated to resist impact, it's intended mostly for prisons and psychiatric hospitals.

TINTED GLASS

Tints can be functional or purely decorative. Some absorb heat and block ultraviolet radiation, thanks to the **metallic oxides (A)** that happen to color them green, gray, blue or bronze. For looks alone, windows can be made any color by **laminating** one or several layers of plastic between sheets of glass **(B).** Tinted glass looks strange in old homes but could be great in some spots in newer ones.

BAD-GUY GLASS

Standard laminated glass resists impact, but it can't withstand bullets or repeated punches. **Secur-Tem + Poly** can. For those of us whose children insist on playing ball indoors, the product, shown here in two thicknesses, is the ultimate in protection. It can be ordered in colors or as a one-way mirror.

Inferno-Lite, FRP 100, 3/16", $40 per sq. ft.
Globe Amerada, 2001 Greenleaf Ave., Elk Grove Village, IL 60007; 800-323-8776.

Secur-Tem + Poly, 5/16", $48 per sq. ft., and 1.4", $58 per sq. ft.
Globe Amerada

Decorative laminated glass: Monsanto Opticolor, $10–35 per sq. ft.
Mirror Factory Inc., 12725 16th Ave. N., Plymouth, MN 55441; 800-452-1644.

LIGHT PIPES

Sun Industries Inc.
Box 887, Bountiful, UT 84011: 800-409-9927.

Solatube
5825 Avenida Encinas, Suite 101, Carlsbad, CA 92008: 800-773-7652.

ODL Inc.
215 E. Roosevelt Ave., Zeeland, MI 49464; 800-253-9000.

Sun Tunnel Sky Lights
786 McGlincey Lane, Campbell, CA 95008; 800-369-3664.

SunPipe Co. Inc.
Box 2223, Northbrook, IL 60065; 800-844-4786.

SunLITE
524 E. Broadway, Logansport, IN 46947; 800-231-1596.

Bronze screening: $2.12 per sq. ft.
Jamestown Distributors, 28 Narragansett Ave., Jamestown, RI 02835; 800-423-0030.

Bronze, copper, stainless steel, brass, etc.
Gerard Daniel & Co. Inc., 5 Plain Ave., New Rochelle, NY 10801; 800-232-3332 or City Wire Cloth, 13900 Orange Ave., Paramount, CA 90723; 310-630-8050.

Wooden screen and storm doors
Air-Tite Manufacturing 33519 SR2 Sultan, WA 98294; 206-793-3435.

light pipes

Light pipes are conduits that bring outdoor light indoors, through the attic to a floor below. This saves the cost of installing and operating an electrical fixture while adding that most precious illumination—sunlight. Most light pipes are about 10 inches in diameter, so they fit between rafters for easy installation. Several models are on the market, starting at about $170. Checkpoints: Some makes aren't bug-proof, others aren't airtight. To help prevent condensation where the pipe passes through the attic, wrap the tubing with insulation.

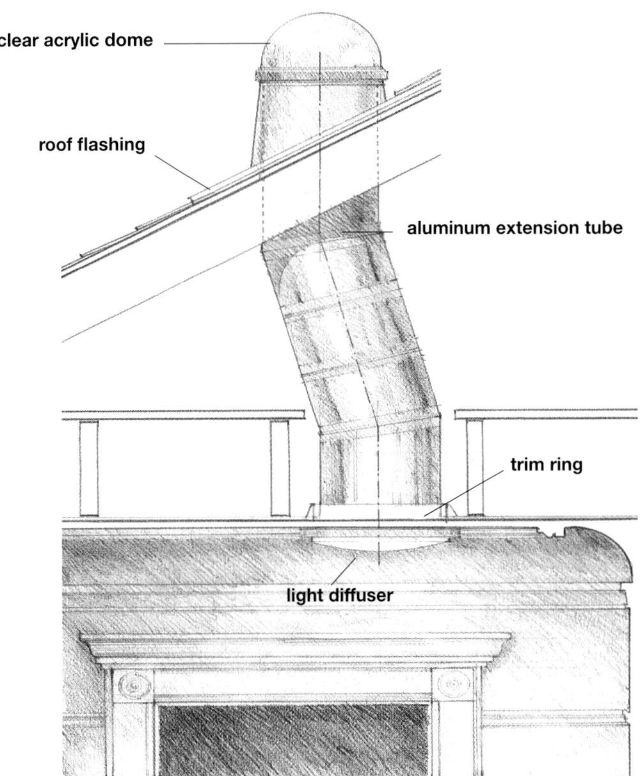

corrosion-resistant screens

Fiberglass screening, the stuff that comes in most new window screens and doors, is the cheapest type; it runs about 25 cents a square foot, is matte black and is impervious to almost anything. If you're willing to pay eight to 10 times more, copper, bronze, brass or stainless-steel screening are excellent, corrosion-resistant options. (Nickel, monel or titanium screens are also available, if price is no object.) Copper is the softest and will, like brass and bronze, acquire a verdigris patina in time. Stainless is the strongest and most costly, and it stays shiny indefinitely. Bronze, the cheapest, shines like gold when new. All have a metallic "flash," which keeps people from running into them accidentally, and they look good. (We've used bronze screening instead of tempered glass panels to childproof a glass-paneled interior door.) The typical mesh size for insect screening is 18 by 14 (18 holes per horizontal inch by 14 holes vertically) with .009 or .011 gauge wire. If a screen must resist bumps from dogs or kids, you can order square mesh (e.g. 14 by 14 or 16 by 16) with thicker wire and a bigger mesh size for better durability. Don't put copper, bronze or brass on aluminum doors; galvanic corrosion will eat at the aluminum.

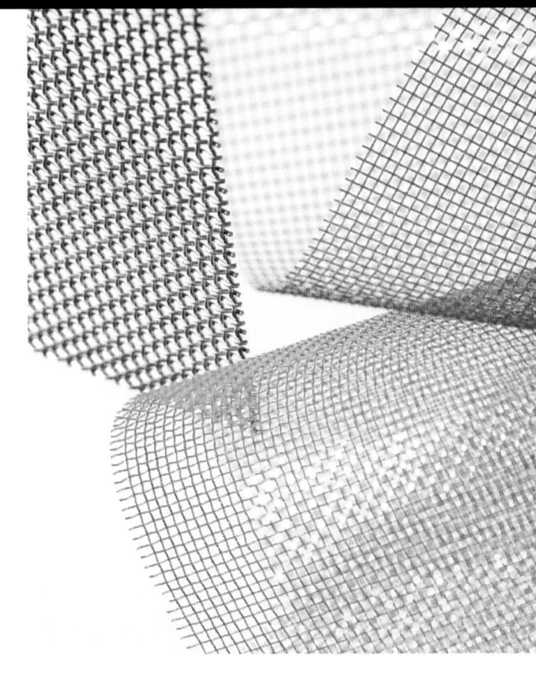

notes

walls and ceilings

DRYWALL | DRYWALL TAPE

VENEER PLASTER | CROWN MOLDINGS | PLASTER MEDALLIONS

drywall

Our eyes, drawn to the fine woodwork, fancy tile and other delights of domestic architecture, overlook 80 percent of almost every room. Hidden in plain sight on virtually every wall and ceiling is a sandwich of paper and gypsum that is smooth, fire-resistant and unloved. Call it Sheetrock or wallboard, plasterboard or drywall, it's the modern-day replacement for plaster. And though drywall falls short of plaster's hardness and durability, it's cheap to buy and simple to install. A room can be painted as soon as it's "rocked," but a three-coat, trowel-applied plaster job must first cure for 30 days. Thus drywall is king of wall and ceiling, wearing its crown by financial default.

The United States produces drywall in titanic volumes: 24 billion square feet each year. To meet demand, millions of tons of gypsum rock are extracted annually from quarries and mines in the Midwest, Canada and Mexico. Since the 1970s, drywall manufacturers have also been "mining" coal- and oil-fired power plants, where gypsum is a waste product of air-pollution control.

To turn raw gypsum into wallboard, the rock is crushed, then calcined—cooked at 350 degrees until it turns into a dry powder called plaster of paris or stucco. This plaster is mixed with water and additives to form a slurry that flows onto a continuous sheet of moving paper made from recycled newsprint. A second piece of paper rolls out on top. The resulting sandwich is fed through rollers that press it to uniform thicknesses from a quarter-inch to a full inch in eighth-inch increments. Quarter-inch board is flexible enough to hug curved walls and ceilings, but most walls are clad in half-inch sheets. Thicker dimensions are needed where additional strength or fire resistance is important.

One of drywall's greatest virtues is how easily it can be customized. It comes in lengths up to 16 feet, and because of the growing popularity of high ceilings, manufacturers now offer 54-inch widths. Placed horizontally, two sheets can cover a 9-foot wall. Likewise, gypsum's natural resistance to fire can be increased by embedding glass fibers or vermiculite in the core. Attempts are being made to combat plaster's biggest enemy—water—with silicone additives, wax emulsions and fiberglass or plastic skins. Can these products, marketed as underlayments for shower stalls, exterior walls, and even roofs, hold up? Yes, if the cores don't get wet.

Drywall's other weakness shows on impact. Furniture, fists, even doorknobs that bounce off a plaster wall can poke though half-inch drywall with relative ease. Part of the reason is that the gypsum core is aerated for lightness (a sheet of half-inch drywall weighs 48 pounds, down from 80 in 1900), so screws and nails can penetrate easily. For impact-resistant drywall, companies mix paper fibers in the core, overlay fiberglass mesh or apply thin sheets of Lexan, a clear, practically unbreakable plastic.

Despite all these improvements, drywall remains a remarkably inexpensive commodity—so inexpensive, it's easy to overlook its aesthetic and functional advantages. Even in its glory days, plaster was rarely as smooth as modern drywall or as easy to repair. And putting up drywall is a far more efficient process.

drywall tape

Fiberglass drywall tape works well because the joint compound (which drywallers call "mud") gets a good grip through the perforations. Too bad it's a bit tricky to handle. And too bad easy-handling paper tape has to be glued in place with a separate application of mud. Weary of balancing these pros and cons, drywall contractor (and fireman) Sal Loscuitto took the best of both and came up with Easy Joint Tape. Perforated paper with a self-stick backing, it eliminates the gluing step and also provides a "soft" stick, so the tape can be repositioned if necessary. It costs significantly more than conventional tapes (about $7 per 200-foot roll), so you have to decide what you want to save: time or money.

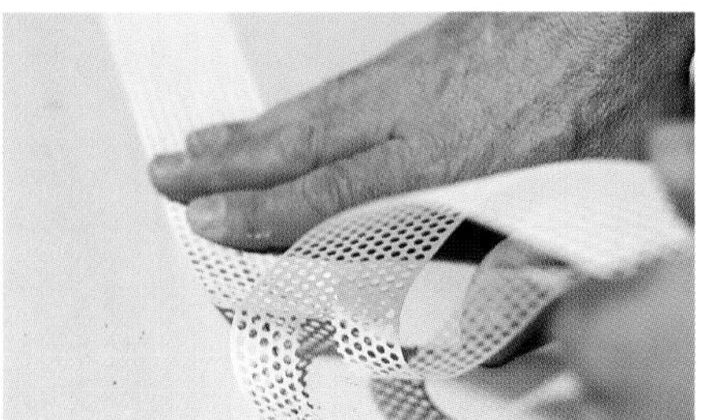

SOURCES

Easy Joint Tape: 100-ft. roll, $3.75-4.25, 200-ft., $6.50-7.50
Wall Tool & Tape Corp., 81-11 101st Ave., Ozone Park, NY 11416; 718-641-6813, fax 718-641-6758.

Blueboard jumbo baseboard, ½"x4'x8', $4.50
Celotex, 4010 Boy Scout Blvd., Tampa, FL 33607; 813-873-1700.

Impact-resistant Lexan 2000, ⅝"x4'x8', $32
National Gypsum Co. Gold Bond Building Products, 2001 Rexford Rd., Charlotte, NC 28211; 704-365-7300.

Impact-resistant VHI Fiber Bond, ⅝"x4'x8', $20
Louisiana Pacific Corp., 111 SW 5th Ave., Portland, OR 97204-3601; 503-221-0800.

Exterior Dens-Glass Gold sheathing, ½"x4'x8', $11
Georgia-Pacific, Gypsum Div., 133 Peachtree St. NE, Atlanta, GA 30303; 313-225-6119.

Moisture-Guard Gyproc, ½"x4'x8', $6-7
Georgia-Pacific, Gypsum Div.

Vapor barrier Foil-Back, ¾"x4'x8', $9
National Gypsum Co.

Fire-resistant Ultra-Code Core, ¾"x4'x8', $13
US Gypsum Co., 125 S. Franklin St., Box 806278, Chicago, IL 60680-4124; 800-851-8501.

Embossed raised panel, $2 per sq. ft.
Pittcon Industries Inc., 6490 Rhode Island Ave., Riverdale, MD 20737; 301-927-1000.

FOR FURTHER READING

Twentieth Century Building Materials, ed. by Thomas C. Jester, 1995, 352 pp., $55
McGraw Hill Co., 1221 Ave. of the Americas, New York, NY 10020; 800-722-4726.

SPECIALTY DRYWALL

SMOOTH

Blueboard has a special paper, tinted for easy identification, that bonds tightly to veneer plaster.

EXTERIOR

Drywall for outdoor use requires advanced technology to protect the core from the elements. This substrate for synthetic stucco has a silicone-treated core and faces of yellow alkali-resistant fiberglass. Comes with a 5-year warranty.

IMPACT-RESISTANT

These boards withstand abuse that would cause regular drywall to crumble. The top panel has a thick layer of pressed paper fiber covering a gypsum core filled with more paper fiber and perlite. Green fiberglass mesh on the back provides additional reinforcement. The bottom board, faced with a 20-mil-thick sheet of Lexan, is 17 times stronger than conventional drywall.

VAPOR BARRIER

In regions with cold, wet winters, aluminum-backed drywall stops moisture trying to seep through walls and into wall cavities. The foil side always faces the inside of the cavity.

MOISTURE-RESISTANT

Water wrecks wallboard. So for walls and ceilings in bathrooms, builders put up this moisture-resistant drywall with green-tinted paper. The gypsum contains water-repellent wax emulsion; the paper is also specially treated. Though designed as an underlayment for wall tile, greenboard shouldn't be used in showers or anyplace that receives more than the occasional splash.

FIRE-RESISTANT

Regular drywall stands up well to fire but loses strength as heat drives out the water in the plaster. (Gypsum is 20 percent water.) This ¾-inch-thick board—rated to withstand a 1,700-degree fire for two hours—contains additives such as glass fibers and vermiculite that hold the panel together as the gypsum turns to dust.

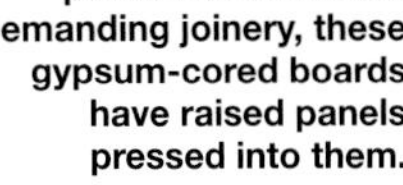

EMBOSSED

For the look of an elegant raised wood panel without all the demanding joinery, these gypsum-cored boards have raised panels pressed into them.

veneer plaster

For all its advantages, drywall has a few drawbacks: It's subject to nail pops, telegraphing joints and abuse (you can scratch it with your fingernail). There is an alternative—veneer plastering—that marries plaster's hardness, fire resistance and texture with the convenience of drywall. Veneer plaster is skimmed over sheets of blueboard, a blue-tinted gypsum wallboard treated to accept plaster. Within an hour, the plaster hardens into a monolithic surface with a perfectly smooth finish.

PLASTERING TECHNIQUE

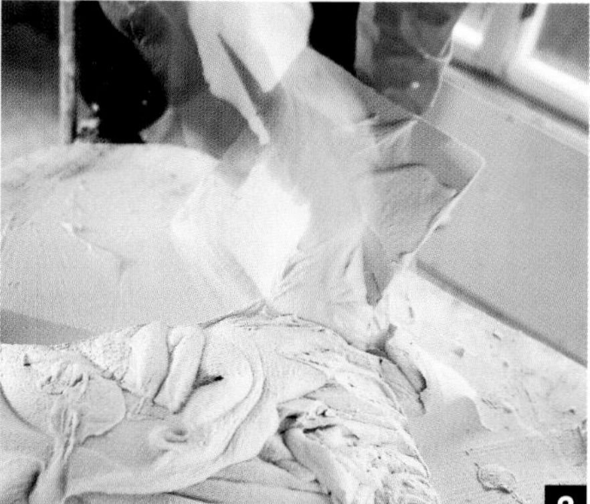

1. After adding a tablespoon of retarder to 6 qts. of water, plasterer John Marshall blends 70 lbs. of an 80-lb. bag of base-coat plaster with a jiffler mixer attached to a 500-rpm drill. Mixing plaster is dusty, hence the mask. Glasses keep caustic materials out of eyes.

2. In about a minute, the batch reaches the consistency of grainy cake batter. Marshall stops mixing immediately; any longer can accelerate the set. With his trowel, he scoops plaster off the mixing board and onto his hawk, which holds plaster as he works.

3. Ceilings are plastered first, then walls. Joints and screws get an initial coat, then the blueboard's entire surface is covered with a layer of plaster about ⅛-inch thick. All tools are thoroughly cleaned with water between each batch.

4. Once the ceiling sets, base-coat plaster is troweled on the walls with a quick up-and-down rhythm. Holding the trowel at about 30 degrees to the blueboard gives a smooth, chatter-free surface, leveling undulating walls and creating perfect intersections.

5. After about an hour, the base coat will have set. Then a finish coat made of lime putty mixed with gauging plaster is spread over the base-coat veneer. The corner trowel Marshall is using puts a neat crease down an inside corner.

6. This second veneer must be "packed" and polished with a trowel to make the smooth, hard surface we know as plaster. No sanding is needed. The wet felt brush in Marshall's left hand moistens the finish coat as he pulls the trowel at a low angle down the wall.

HOW THE PROS DO IT

Checking for level: "Plaster's greatest asset is as a leveling device," says Jim Marshall, John's father. "You don't have to fuss with shimming out boards; the plasterer takes care of it." Marshall uses 4- and 8-foot wooden straightedges to find dips in the blueboard, then circles those spots with pencil.

Working time: Heat, humidity and drafts speed the set, as will bits of plaster left over from a previous batch. That's why Marshall keeps the thermostat turned down, the windows shut and is fastidious about cleaning his tools and mixing board.

Troweling technique: Always work plaster from wet areas into dry, Marshall recommends. For smoothing, the trowel is pressed down slightly at the front to feed the plaster off the back. Hold the trowel at a low angle (30 degrees or less); a 45-degree angle roughs up plaster, leaving "chatter marks."

Creating texture: Many of Marshall's clients ask him to leave a little roughness or irregularity in texture so visitors won't mistake his work for drywall. What he can't abide, however, is seeing his finely polished, perfectly smooth plaster surfaces painted with a thick-napped roller. "It leaves an orange-peel surface that looks like painted drywall," he says. Use a short-napped roller or, better yet, a brush, and "your walls will look like a piece of furniture."

Squaring corners: A straight, square corner is the mark of a good plaster job. In Marshall's hands, all it takes is a couple of swipes with the edge of his 3-foot slicker, a simple beveled cedar clapboard, followed by a clean cut with his corner trowel.

THE MATERIALS

Clean water (1) and fresh plaster (2) are essential. "Look for the date on the bag," advises Marshall. "If it's more than six or eight months old, pass it by." He buys bags with the same date for consistency. Retarder (3) and accelerator (4) slow down or speed up the set. The jiffler mixer (5) and the low-speed drill (6) make quick work of mixing. Gauging plaster (7) and hydrated dolomitic lime (8) form the finish coat. Lime is caustic; cover skin, hands and eyes, and wear a dust mask when mixing.

THE TOOLS

Ninety percent of the time, the only tools a plasterer uses are a hawk (1) and steel trowel (2). The hawk holds the plaster and gives a convenient edge for scraping the trowel clean. The pipe trowel (3), tuck trowel (4) and margin trowel (5) are handy when space is tight. The point trowel (6) cleans up blobs at the edges of outlet boxes. The corner float (7) and corner trowel or "butterfly" (8) smooth inside corners. A felt brush (9) smooths surface imperfections. Nonetheless, Marshall says, "A plasterer's best tools are his arm and his eyes."

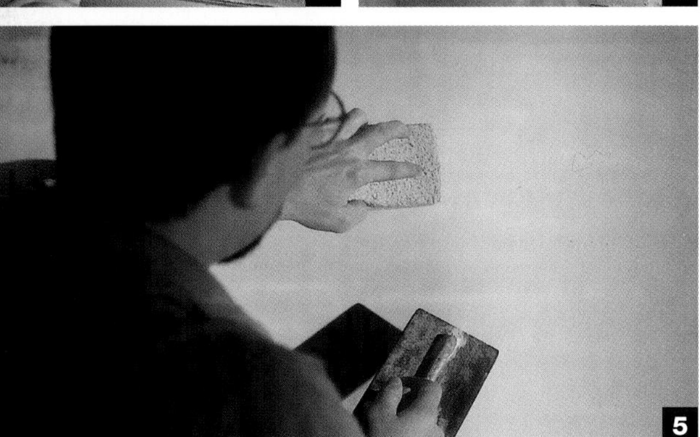

PATCHING PLASTER

While it may be tempting to rip down an old plaster-over-lath wall and replace it with drywall, often the plaster can be saved. Firm plaster that has popped off lath can be reattached with plaster washers and drywall screws. Soft, crumbly plaster should be removed down to the lath and out as far as the firmly attached areas in preparation for patching.

John Marshall starts by chiseling out the old plaster from between the lath and brushing away any loose dust and plaster crumbs. The plaster around the hole is scraped smooth with an angle plane or rough drywall screen.

1. He brushes a bonding agent over the lath and all plaster being recoated, then covers all but the smallest cracks with fiberglass mesh tape. To make lime putty for the patch, Marshall mixes hydrated dolomitic lime with water until it is the consistency of whole-fat yogurt. He then slakes it for at least 20 minutes, leaving a thin layer of water on the surface to ensure even rehydration. The putty is formed into a ring on the mixing board and the bonding agent is poured into its center.

2. Marshall sifts in several handfuls of gauging plaster to give the putty more body.

3. He mixes everything with his trowel into a stiff dough, which he presses firmly onto the exposed lath and around the edges of the hole, leaving a slight depression for the final coat.

4. To reinforce the patch, he cuts a sheet of fiberglass mesh to cover the hole and presses it into the wet plaster.

5. After the first coat sets, Marshall mixes another batch of lime putty and gauging plaster (this time with less gauging so it's easier to work). He then skims a thin final coat over the entire area and uses a sponge to touch up any surface imperfections. In 30 minutes the patch has set and the job is finished, without any need for sanding.

crown moldings

Ever since Renaissance architects borrowed the idea from Greek and Roman temples, crown molding has been an elegant way to cover the hard-edged joint between wall and ceiling. Its graceful, ogeed fillip—a ripple on the boundary of horizontal and vertical—frames a ceiling, elevating it to a higher plane.

The twists and turns of a seamless crown are admirable but tough to put together. Although the molding in situ presents the illusion of being solid, it is usually made of thin strips of wood, which bend and sag when they are being installed. Unless carefully measured and cut, there are likely to be ugly gaps, ill-fitting joints, misaligned profiles and a large pile of scrap.

Carpenters who install crown molding are more like cabinetmakers than framers. When a trim carpenter messes up, his mistakes are on permanent display. A framer's errors get buried under drywall.

Bernie diBenedetto knows he cannot afford to make mistakes. After eight years of installing trim and building cabinets, he works with a single-minded concentration that defies interruption, patiently shaving here and there until only a razor-thin line shows where two pieces meet. Stained moldings demand such fussiness: Gaps can't be covered with caulk and paint. But even if the wood will be painted, diBenedetto works to a stain-grade standard. As *This Old House* painting contractor John Dee says, "You don't need much caulk on Bernie's work."

Tight joints are the measure of quality workmanship, but materials count too. DiBenedetto handpicks his stock to weed out burn marks, sap pockets and any other defects. He also makes sure all profiles are identical. If two pieces don't match exactly, almost nothing can be done to make a joint fit.

SOURCES

Accelerator: carton of eight 2-lb. bags, $17; Imperial base-coat plaster: 80-lb. bag, $14; Imperial blueboard: type 1px1, ½", $8 per 4"x12" sheet; ⅝" fire code, $11.28 per 4"x12" sheet; Structo-Gauge gauging plaster: 100-lb. bag, $26; Redtop retarder: 24 oz., $4
US Gypsum Co., 125 S. Franklin St., Box 806278, Chicago, IL 60680-4124; 800-874-4968.

Aluminum hawk: #5, 12"x12", $14.26; Angle float: #5766, 10"x4½"x⅞", $11.22; Xtralite curved-handle trowel: #MXS7, 12"x5", $24.49; Drywall corner trowel: #23, 4"x5", $10.98; Jiffler mixer: #893, $17.20; Pointing trowel: #45-S, 5"x2½", $9.68
Marshalltown Trowel Co., Box 738, Marshalltown, IA 50158; 800-888-0127.

White fiberglass mesh tape: #84389-15774, 2½", $9.53; Blister brush: #84389-03224, $25.14
Goldblatt Trowel, Trade Tools Div. Stanley-Proto Industrial Tools, 14117 Industrial Park Blvd. NE, Covington, GA 30209; 404-787-3800.

Ivory finish lime: 50 lbs., $8.99
Gem-lime Group, Box 158, Genoa, OH 43430; 800-537-4489.

Low-speed drill: #1101-1, single-speed, triple-gear reduction Hole Shooter for ½" drill bits, 500 rpm, $292
Milwaukee Electric Tool Co., 13135 W. Lisbon Rd., Brookfield, WI 53005; 800-274-9804.

FOR MORE INFORMATION

Plastering Skills, by F. Van Den Branden and Thomas Hartsell, 1984, 543 pp., $32
American Technical Publishers Inc., 1155 W. 175 St., Homewood, IL 60430; 708-957-1100.

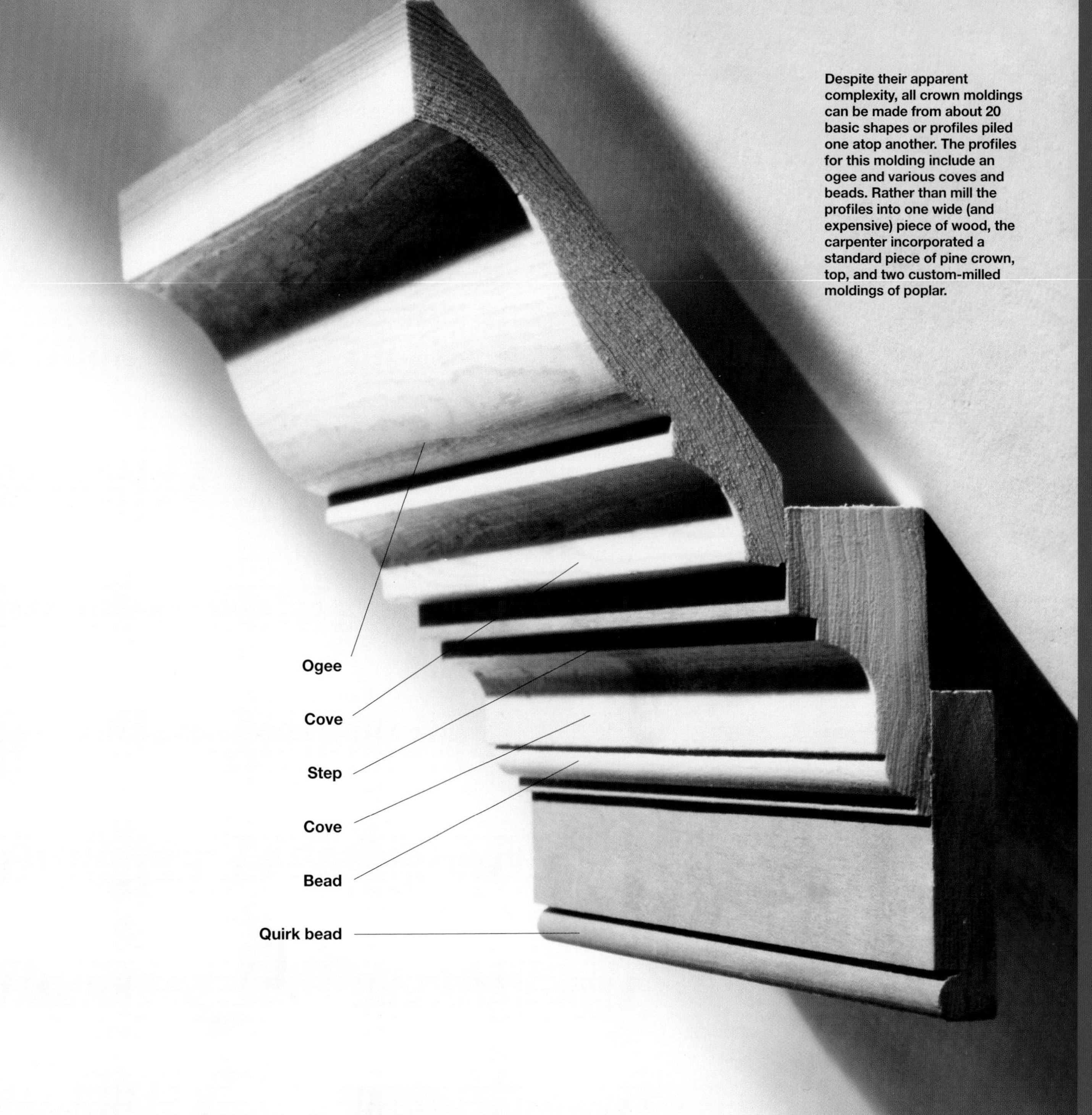

Despite their apparent complexity, all crown moldings can be made from about 20 basic shapes or profiles piled one atop another. The profiles for this molding include an ogee and various coves and beads. Rather than mill the profiles into one wide (and expensive) piece of wood, the carpenter incorporated a standard piece of pine crown, top, and two custom-milled moldings of poplar.

To install crown molding, a trim carpenter has to know how to cut three types of joints: scarf, miter and cope. A scarf is an overlapping joint between two pieces in a straight run of molding. If diBenedetto had his choice, a molding would always be long enough to reach from corner to corner without a scarf. When it's not, he cuts pieces at a 22½-degree angle, then glues and nails them. For large moldings, which can be quite hard to align, he makes a straight cut and joins sections with biscuits.

Miter joints are used wherever two pieces of molding meet at an outside corner, like the corners on the outside of a box. Miters are tough because cuts must line up exactly, and even the most perfect cuts are thrown off if the walls and ceiling aren't square. Good trim carpenters leave pieces a bit long and shave them down with a miter saw or block plane until they fit. A "one-cut carpenter" is diBenedetto's code name for someone who does sloppy work.

On inside corners (like those on the inside of a box), diBenedetto uses the more forgiving cope joint. A cope is a cut in the end of one molding that follows the contour of the profile. The coped piece then overlaps its neighbor.

When installing molding in a room with only inside corners, diBenedetto's strategy is to install the first piece on the wall opposite the main doorway, each end cut square and butted into the corners. He copes one end of each of the side-wall pieces and makes straight, 90-degree cuts on the ends that butt the opposite corners. All measurements are taken corner to corner, in line with the bottom edge of the molding. The last piece over the doorway is coped on both ends.

BUYING THE RIGHT AMOUNT

Half the trick to using crown molding efficiently is making accurate cuts. The other half is buying no more than needed. To calculate how much to buy, sketch out the room's perimeter, then mark the length of each wall, indent and projection to which molding will be added. Arrange the pieces on paper so as to have as few joints as possible. Most moldings come in 16-foot lengths, so a 10-foot wall in one part of the room and a 4-foot-deep alcove in another can be cut from the same piece. At every joint, add 2 inches to each piece of molding to allow for cutting. Divide the total number of feet by 16 (or the longest length the molding comes in), then round up to the nearest whole number. If two pieces of molding have to be joined in the middle of a run—when a wall is more than 16 feet long, for example—try to place the joint in an inconspicuous spot, such as over a door, not directly opposite the entry where it will be the first thing people see.

FIXING GAPS

Crown molding can't beautify a room if it draws attention to the gaps in an uneven ceiling. If the gaps are small, try pushing the top of the molding up and back toward the wall. If that doesn't work, fill the gaps with caulk. Small bumps can be accommodated by shaving off the molding's top edge with a block plane. (Don't try this near outside corners.) There's a limit to how much a molding can be bent without throwing off the joints or planed without anyone noticing. When facing a ceiling with large bumps and hollows (and a budget that doesn't allow for a plasterer to level it), try leaving a slight gap between the molding and the ceiling—a quarter of an inch will do nicely. Secure the crown with triangular nailing blocks.

COPING WITH A TRICKY CUT

1. The first step in coping a crown molding is to reveal its profile by cutting a 45-degree inside miter. Crown molding is always cut face up and upside down, with its top edge on the saw's base and bottom against the fence. For safety and best results, lower the blade slowly through the work and let it come to a stop when the cut is complete. A spinning blade can tear splinters from the cut or snag the wood—not to mention a finger.

2. After the inside miter is cut, it's time to cut the cope. DiBenedetto highlights the edge of the profile with a pencil, then follows the pencil line as he saws off the miter's exposed end grain. What's left is a mirror image of the profile. Angling the saw so it undercuts the profile helps ensure a tight fit. DiBenedetto cuts most copes with an electric jigsaw and a 14-teeth-per-inch blade, but the tool has to be used with great care or the wood will splinter. Here he's using a coping saw, the method Norm Abram prefers.

3. After making the cope, test it. If the molding doesn't fit, take more off the back side with a rasp, as shown here, or utility knife.

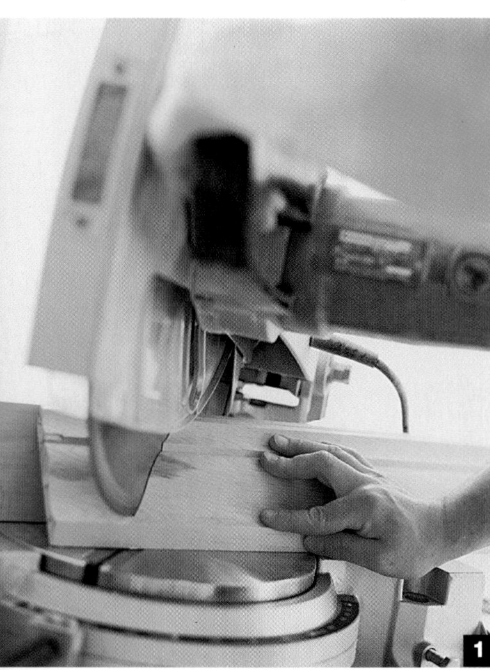
1

2

3

Crown molding: B8013 pine crown, $2.27 per lineal ft.; custom poplar Hartwright crown, $1.75 per lineal ft. for 250 feet; custom poplar pencil bead crown, $1.38 per lineal ft. for 250 feet

Concord Lumber Corp., 55 White St., Littleton, MA 01460; 508-369-3640.

Miter saw: C15FB Hitachi 15", $1,346

Hitachi Koki USA, 3950 Steve Reynolds Blvd., Norcross, GA 30093; 800-706-7337.

Coping saw: Disston #15, $6

Greenfield Ind., Disston Div., Deerfield Industrial Park, S. Deerfield, MA 01373; 800-446-8890.

File: Nicholson 8" four-in-hand rasp, $9.82

Cooper Tools, 1000 Lufkin Rd., Apex, NC 27502; 919-781-7200.

Air-powered brad nailer: Airy #8TK0241, $109-119

Airy Sales Corp., 1425 S. Allec St., Anaheim, CA 92805; 800-999-9195.

MAIL-ORDER MEDALLIONS

Free catalog

Felber Ornamental Plastering Corp., Box 57, 1000 W. Washington St., Norristown, PA 10404; 800-392-6896.

Catalog #130, $4

The Decorator's Supply Corporation, 3610 S. Morgan St., Chicago, IL 60609; 312-847-6300.

PERFECTING AN OUTSIDE MITER

1. One secret of a tight miter joint is keeping the pieces in precise alignment. To do this, diBenedetto puts marks on the wall and ceiling to help guide crown placement before he hammers. After preparing the base moldings, he uses a square to mark where the crown's bottom edge will rest. **2.** Then, with a scrap of molding held to this mark, he makes matching marks on the ceiling.

3. An air-powered brad nailer speeds a trim carpenter's work. There's no pounding to jostle joints, it's one-handed, and the fastener sets automatically below the surface with just a pull of the trigger. The brads only tack the molding in place. Finish nails will be hammered into the framing later; these should be long enough to penetrate as far as the molding is thick. It's difficult to get miter joints tight and keep them that way. **4.** DiBenedetto finds out-of-square corners—the bane of trim carpenters—by holding a mocked-up sample against all outside corners. If the sample doesn't fit, he knows he'll need to spend more time trimming the real thing. This test fit of mitered molding shows a slight gap near the top. Wood glue brushed on the miter cuts and a 4d nail shot through the top of both pieces will close up the gap. Once everything's tight, diBenedetto cleans up the edge with 120-grit sandpaper.

plaster medallions

While plaster has covered walls since the days of the pyramids, Greeks in the fifth century BC were the first to mold plaster for ornament. The Romans adopted the craft, but it died when the empire dissolved. Then, the story goes, a 13th-century Italian monk accidentally spilled plaster on a marble statue and rediscovered the material's marvelous molding properties. By the 15th century, at the height of the Italian Renaissance, plaster craftsmen known as stuccotori were busily filling orders for classical decoration. Little has changed since. Brushable rubber has replaced animal-hide gelatin for molds; fiberglass reinforcing mesh can substitute for burlap, sisal or goat hair; and polyvinyl acetate is added to improve adhesion. But the techniques, and the gypsum plaster itself, are no different from those used in Donatello's time. To make your own renovations easier—and to avoid doing complicated plaster work yourself or hiring an ornamental plaster artisan—consider using one of many premolded mail-order ceiling medallions for your finishing touch.

kitchens and baths

COUNTERTOPS | MARBLE COUNTERTOPS

KITCHEN FLOORING | KITCHEN SINKS | SHOWERHEADS AND FITTINGS

LOW-FLOW TOILETS | REPLACING A TOILET | TILING A BATHROOM

countertops

1
STAINLESS
STEEL

2
MAN-MADE
SLAB

3
HIGH-
PRESSURE
LAMINATE

4
STONE
SLAB

5
CONCRETE

Countertops are the most visual and tactile kitchen elements. It's easy to fall in love with one type to the exclusion of others. Don't. In his home, *This Old House* host Steve Thomas disregarded common sense and set his sink into a maple counter. The wood's expansion and contraction with changes in humidity repeatedly breaks the polyurethane sealant he used on the sink rim. Despite diligent mopping, the wood is now cracking.

It's smart to put different sorts of countertops in locations that make the best use of their strengths. Ceramic or stone tile can go next to stoves so hot pans can be placed there. Butcher block can be set into an island surface meant for slicing and chopping. Laminate or solid surfacing can be used everywhere else. Another solution is pullouts. Steve likes a kitchen he saw with lavish polished granite counters and no fewer than six pull-out cutting boards at key points around the room.

Countertops that meet a wall commonly have a backsplash—a vertical section about 4 inches high to guard against spills. But it makes sense to go higher. And there's no rule that backsplashes must be made of the same material as counters. One of Steve's favorite small kitchens, squeezed into an urban closet, uses mirrors as backsplashes to expand the sense of space.

1. STAINLESS STEEL

INSTALLATION: Glued to plywood substrate. Professional installers form edges and backsplash.
PROS: Stain-proof, scorch-proof, difficult to dent, industrial appearance.
CONS: Fabrication is expensive; industrial appearance.
PRICE: $100 per lineal foot, installed.

2. MAN-MADE SLAB

INSTALLATION: Screwed in place or secured with silicone adhesive.
PROS: Very durable. Resists heat, scorching, etching, staining.
CONS: Unusual in homes; installers may be inexperienced. May appear more commercial than residential.
PRICE: $23 per square foot, plus installation.

3. HIGH-PRESSURE LAMINATE (FORMICA, WILSONART, MICARTA)

INSTALLATION: Sheets are glued to medium- or high-density fiberboard backing. Ready-made lengths available at home centers.
PROS: Inexpensive, widely available; tough surface; wide range of colors, patterns and textures; easy to clean; well-suited to most cooking tasks.
CONS: Hard to repair; not suitable for cutting; brown edge may show at seams; can be scorched by hot pots.
PRICE: $20–45 per lineal foot, installed.

4. STONE SLAB (GRANITE OR MARBLE)

INSTALLATION: Same as man-made slab, but base cabinets may have to be reinforced to support extra weight.
PROS: Long-lived; luxurious in appearance; impervious to scorching; good for baking areas.
CONS: Expensive; hard; cold; marble may etch or stain if improperly sealed.
PRICE: $150 and up per lineal foot, installed.

5. CONCRETE

INSTALLATION: Poured into molds, removed and installed like stone slabs.
PROS: Scorch-proof; hard to scratch; looks like stone slab at lower cost.
CONS: Edges can chip; must be sealed regularly.
PRICE: $75–200 per lineal foot, installed.

SOURCES

NOTE: IN COUNTERTOP LISTINGS, ONE LINEAL FOOT IS A 25-INCH-DEEP SECTION OF COUNTER 1 FOOT WIDE.

Stainless steel: $75 per lineal foot, for fabrication only; installation can run to $1,000 for 8 hours of work by a two-man crew
Modern Stainless, 33395 Railroad, Union City, CA 94587; 510-471-9191.

Man-made slab: Fireslate; $23 per sq. ft.
Fireslate–2 Inc., 47 Hamel Rd., Lewiston, ME 04240; 800-523-5902.

High-pressure laminate: Formica in Saffron 7024-58; $3–3.50 per lineal foot for laminate only; $12–15 per lineal foot, installed
Formica Corp., 800-367-6422.

Stone slab: Verde Maratonka granite, 1½" slabs, $45 per sq. ft.; Travertine marble, 1½" slabs, $35 per sq. ft.; Premier bluestone, 1¼" slabs, $12.05 per sq. ft. (additional cost for all edge finishing)
Bergen Bluestone, 404 Route 17, Paramus, NJ 07652; 201-261-1903.

Granite
Creative Stone Accessories Inc., 104 E. Central Junction Drive, Central Industrial Park, Savannah, GA 31405; 912-234-8485.

Concrete: Custom fabricated pigmented concrete countertop; $85 per sq. ft., plus shipping and freight costs from the west coast
Ann Sacks Tile & Stone Inc., 5 E. 16th St., New York, NY 10003; 212-463-8400.

Wood: Butcher-block slab fabricated and finished as countertop
David N. Johnson Cabinetmaker, 620 W. 47th St., New York, NY 10036; 212-423-3711

Soapstone: Approx. $140 per running foot (25" depth) installed
Vermont Soapstone Co., Box 168, Stoughton Pond Rd., Perkinsville, VT 05151; 802-263-5404.

Exotic metals: copper, $125 per lineal foot; zinc, $45 per lineal foot; both for fabrication only
Modern Stainless.

6. WOOD (SUCH AS BUTCHER BLOCK)

INSTALLATION: Screwed in place or secured with silicone adhesive.
PROS: Natural warmth; resilient surface; damage can be sanded away; good for chopping.
CONS: Susceptible to warps and cracks, stains, water damage and bacteria buildup (unless disinfected regularly); should not be used near sinks.
PRICE: $50–60 per lineal foot, installed.

7. SOAPSTONE

INSTALLATION: Same as stone slab.
PROS: Authentic material for 19th-century-style kitchen; resists heat.
CONS: Experienced installers are rare; stains easily; challenging to maintain; more susceptible to cracking and chipping than harder stone.
PRICE: $140 per lineal foot, installed.

8. EXOTIC METALS (ZINC, COPPER, RUSTED STEEL)

INSTALLATION: Same as stainless steel.
PROS: Durable; scorch-proof.
CONS: Must be thoroughly waxed and sealed for food safety. Zinc stains badly and is recommended for bars only, not food preparation, because zinc layer is only 1/32 inch over steel.
PRICE: $60 and up per lineal foot, installed.

9. CERAMIC TILE

INSTALLATION: Should be laid on cement backer board or in a bed of mortar, not glued directly to plywood or particleboard.
PROS: Inexpensive; durable; resists heat and scratches; easy to clean.
CONS: Professional installation adds to cost; can chip; grout may discolor; requires sealing; uneven surface makes baking tasks difficult.
PRICE: $5–25 per square foot.

10. SOLID SURFACING (AVONITE, CORIAN, SURELL)

INSTALLATION: Fabricated by factory-licensed shops.
PROS: Seamless; very durable; easy to clean and repair; resists scorching; sinks of the same material can be glued to underside of counter, giving seamless bond.
CONS: Expensive; makers discourage do-it-yourselfers.
PRICE: $100–200 per lineal foot, installed.

TIP

When designing a kitchen, keep in mind the practical implications of the style you choose. Informal styles, like the country look, can tolerate a good deal of clutter. Sleeker or more formal styles require much more diligent housekeeping to look presentable. Be realistic about the style that will best suit you.

marble countertops

VARIETIES AND FINISHES

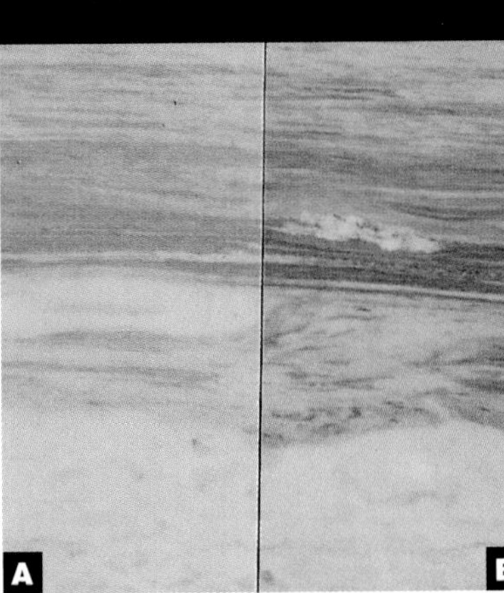

White marbles such as Danby Imperial, Colorado Yule, Carrara or travertine begin as limestone or calcite; black or green marbles are metamorphic basalt or forms of magnesium silicate called serpentine and olivine. Choosing one is largely a matter of aesthetic preference, although some professionals recommend staying away from white marbles in kitchens because of their propensity for staining. Silicate marbles, which are typically darker, are less reactive with acid, though bleaching is more obvious. Highly veined or variegated marbles conceal stains more easily.

Once you've settled on a type of marble, it's important to specify a finish. The vast majority of residential marble is polished. We prefer a honed finish (example A), which is less reflective, easier to maintain and looks softer than a polished surface (example B). Honed marble is finished to a 180 to 600 grit, while polished marble may be finished to an 800 to 3,000 grit and then buffed with oxalic acid until it takes on a mirror-like sheen.

"Honing is the traditional finish for marble," says Martin Hemm of Carl Schilling Stone Works in Proctor, Vermont. "For some reason highly polished finishes became popular in the 1960s and '70s, but they don't look as good and fortunately are losing favor. Polishing makes no sense. It makes it difficult to see the natural luster of marble." Nor does polishing ensure durability; a honed finish resists stains just as well. And polished marble can require greater upkeep. If scratched or etched, it will need professional restoration to return it to its original gloss.

Since much marble arrives at the retailer already polished, you may not be asked to choose. Don't worry: If you want a honed finish, it is a simple matter to return the stone to the polishing bed and tone down the shine.

When selecting unpolished marble, wet the surface of the slab with a sponge to reveal veining patterns and color. Large veins and other imperfections are part of the beauty. Dealers often send customers to the wholesale stone yard to select their slab; you should insist on it. Expect to pay about $500 to $750 for a 30-by-72-inch tabletop; the average installed price for a kitchen counter with backsplash is $100 per lineal foot, though prices vary greatly depending on the type of stone.

STANDARD DIMENSIONS

Domestic slabs of marble come in ⅞-inch and 1¼-inch thicknesses (¾-inch is typical for imports). Lengths of 8 to 10 feet are standard; longer than 12 feet requires a custom quarry request, which can run into the thousands of dollars.

CARE AND CLEANING

Acid is the enemy of marble. It will etch any marble surface and remove the finish on polished stone. Use mild, pH-neutral cleaners like Murphy's Oil Soap or Ivory liquid for routine cleaning. Keep vinegar and citrus fruits and their juices away from marble countertops. "Always use a cutting board," cautions Jonathan Zanger of Westchester Marble and Granite, one of the nation's foremost wholesalers of natural stone, in Mount Vernon, New York. "Marble is very porous and stains easily. The good news is that whatever goes in will come out," Zanger says. "It's like an open door." Remove water-borne stains with hydrogen peroxide. For tougher stains use a poultice available from stone dealers. To minimize absorption, Zanger recommends sealing marble with a nontoxic penetrating sealer like Miracle Sealant's Porous Plus or 511 Impregnator. Avoid sealers that just coat the surface and do not penetrate.

SOURCES

Ceramic tile: Top, LS Obidos in blue, 5⅝" x 5⅝"; Bottom, RO Bordeaux in red, 5⅛" x 5⅛"
Country Floors, 15 E. 16th St., New York, NY 10003; 212-627-8300.

Solid surfacing: Formica Surell in Pinedust; $80–120 per lineal foot (25" depth), installed
Formica Corp. 800-367-6422.

Marble
Carl Schilling Stoneworks, 62 Main Street, Proctor, VT 05765; 802-459-2200.
Rutland Marble & Granite Co., Box 807, Castleton, VT 05735; 802-468-5636.

511 Impregnator sealant for polished stone, $95 per gal.; 511 Porous Plus for absorbent stone, $225 per gal.
Miracle Sealants, 12806 Schabarum Ave., Building A, Irwindale, CA 91706; 800-350-1901.

FOR MORE INFORMATION

***"Care & Cleaning for Natural Stone Surfaces,"* 1995, 16 pp., free**
Marble Institute of America, 30 Eden Alley, Suite 201, Columbus, OH 43215; 614-228-6194.

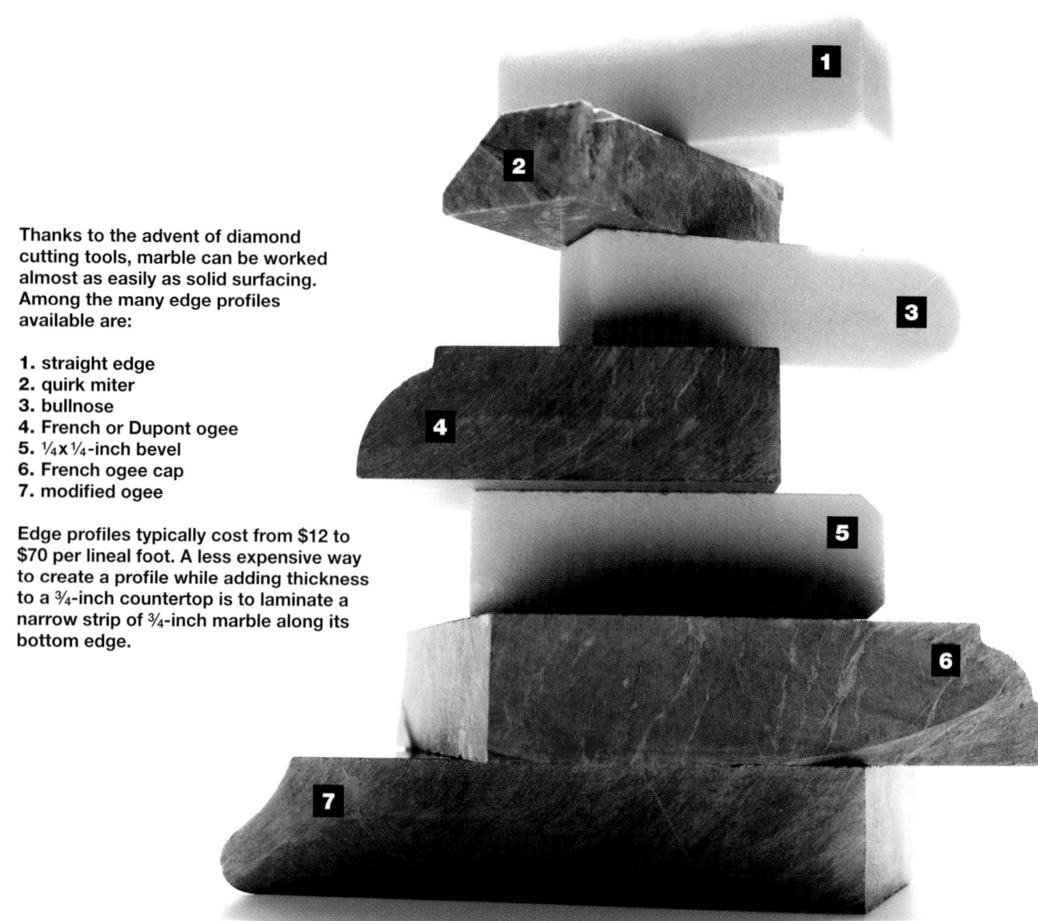

Thanks to the advent of diamond cutting tools, marble can be worked almost as easily as solid surfacing. Among the many edge profiles available are:

1. straight edge
2. quirk miter
3. bullnose
4. French or Dupont ogee
5. ¼x¼-inch bevel
6. French ogee cap
7. modified ogee

Edge profiles typically cost from $12 to $70 per lineal foot. A less expensive way to create a profile while adding thickness to a ¾-inch countertop is to laminate a narrow strip of ¾-inch marble along its bottom edge.

kitchen flooring

Vinyl tile: Classic Touch Grand Mosaic in Raintree/Teal (GR-98), 12" x 12" x 1/8"; $3.50 per tile
Congoleum Corp., 3705 Quakerbridge Rd., Box 3127, Mercerville, NJ 08619; 609-584-3000.

Hardwood: Sterling Strip oak flooring in Natural (C-720); 2 1/4" wide by 3/4" thick in varying lengths; approx. $6 per sq. ft.
Bruce Hardwood Floors, 16803 Dallas Parkway, Dallas, TX 75248; 800-722-4647.

Ceramic and stone tile: San Miniato Relief terra-cotta tile, 10" x 10"
Country Floors, 15 E. 16th St., New York, NY 10003; 212-627-8300.

Linoleum: Marmoleum Dual in Calico (#713) and Firedance (#775), approx. $3.50 sq. ft., installed
Forbo Industries Inc., Humboldt Industrial Park, Box 667, Hazelton, PA 18201; 800-233-0475.

Cork tile: Unfinished (#6300) and wax (#200) finishes, 12" x 12" x 3/16", $2.78 and $3.44 per sq. ft., respectively
Dodge-Regupol Inc., 715 Fountain Ave., Box 989, Lancaster, PA 17608; 717-295-3400.

Terrazzo: Neptune style N546 (light gray), $2.64 per sq. ft.; New Rustic N544 (darker tile), $3.05 per sq. ft. Each tile is 11 13/16" square by 3/4" thick
Wausau Tile Inc., Box 1520, Wausau, WI 54402; 800-388-8728.

Rubber: Cerrito Collection Red (#186), $4.41–4.73 per sq. ft.; Cerrito Collection Seafoam (#115), $3.57–3.83 per sq. ft.; both Vantage Profile Raised Circular Design (5/32" high with profile). Slate (#175) Parquet Design (9/64" high with profile), 19 11/16" square, $3.57–3.83 per sq. ft.
Roppe Corp., 1602 Union St., Box X, Fostoria, OH 44830; 800-537-9527.

VINYL TILE

INSTALLATION: Glued to subflooring; tiles butt, edge tiles cut to fit; patterns need careful planning; classic do-it-yourself project.
PROS: Simple to install; easy to repair; wide range of color and pattern options.
CONS: Poorly installed tiles can crack or lift; floor has many seams.
PRICE: $2–10 per square foot, installed.

HARDWOOD

(includes oak, cherry, birch, maple and hard yellow pine; laminated or solid strips, square tiles)
INSTALLATION: Strips are nailed, glued or screwed to subfloor, tiles are glued.
PROS: Warm underfoot; not as hard as stone or tile; polyurethane finishes resist stains and water damage.
CONS: Frequent refinishing may be required; may squeak if badly installed; shows wear; dirt collects in grooves and cracks.
PRICE: $5–15 per square foot, installed.

CERAMIC AND STONE TILE

(includes terra cotta, granite, marble, slate, pigmented concrete, quarry, saltillo and terrazzo tiles)
INSTALLATION: Often set in thin mortar layer on backer board; seams filled with cement- or epoxy-based grout.
PROS: Durable; impervious to dents and scratches.
CONS: Hard surface; cold underfoot; noisy; dirt can collect in grout; can be costly.
PRICE: $5–25 per square foot, on average; some tiles cost $100 per square foot.

LINOLEUM

INSTALLATION: Laid like sheet vinyl, but joined with heat, not chemical adhesives.
PROS: Inherently antiseptic; pleasant linseed oil smell; no man-made ingredients; solvent-free.
CONS: Now a specialty product, so difficult to find.
PRICE: $4–5 per square foot, installed.

CORK TILE

INSTALLATION: Laid like vinyl tile.
PROS: Soft underfoot; good heat and noise insulation; renewable resource.
CONS: Less durable than some options; unfinished cork requires frequent sweeping.
PRICE: $2–6 per square foot, installed.

TERRAZZO

INSTALLATION: Mix of marble chips with cement or resin base poured in place. Cement base requires concrete substrate, epoxy base requires only plywood; surface is ground smooth.
PROS: Very durable; can be sanded or sealed to look like new.
CONS: Installers are rare; installation is messy and expensive.
PRICE: $25 and up per square foot, installed.

RUBBER

INSTALLATION: Laid like vinyl in sheets or tiles.
PROS: Durable; almost maintenance-free; comfortable underfoot; resists spills; many textures available; industrial look.
CONS: Can be cut by sharp objects; industrial look.
PRICE: $5 and up per square foot, installed.

SHEET VINYL

INSTALLATION: Glued to subfloor. Challenging for amateurs.
PROS: Inexpensive; soft underfoot; easy to clean; few seams.
CONS: Slippery if wet; can be cut or sliced by sharp objects; may last only 10 years; subfloor flaws show if poorly installed; smelly when new.
PRICE: $2–10 per square foot, installed.

CARPETING AND AREA RUGS

INSTALLATION: Carpet laid as in other rooms; rugs require nonskid pads.
PROS: Inexpensive; quick to install; soft underfoot; easily replaced.
CONS: Distinctly unhygienic; may retain odors from cooking or spills.
PRICE: $1–4 per square foot, installed.

10. EPOXY

INSTALLATION: Squeegeed or rolled in place by professionals.
PROS: Durable; no seams to collect dirt.
CONS: May be difficult to find experienced installer; abrasive surface.
PRICE: $5–9 per square foot, installed.

kitchen sinks

Not many years ago, redoing a kitchen meant tough decisions about everything except the kitchen sink. The sink was simple: white-enameled cast iron or silvery stainless steel, one bowl or two. Now picking a sink requires more research than selecting a computer. There are farm sinks, pantry sinks, vegetable sinks and half sinks. Sinks for corners, islands, bars and counters. They are mostly round or square, but some are shaped like amoebas. They are gold-plated, solid brass, slate, stone, fireclay and plastic. They are $29 and tinny or $2,300 and solid copper. Some are practically works of art.

SINK OPTIONS

The sinks people select are dictated in part by the accessories available. Wood or polyethylene cutting boards, colanders, drainboards, plastic sink liners, dishpans, plate racks—all may be provided by the manufacturer, usually at an additional cost, to fit the shape of the basin. Most sinks come with three to five predrilled holes in the deck, or back, for the faucet, sprayer and various types of dispensers.

1. Dish racks and cutting boards fit sink contours for neat appearance and convenience.
2. A plastic colander is handy for straining pasta and rinsing fresh vegetables.
3. Wire basket nests securely, won't slip around in use. Protects sink surface from scratches.

SOURCES

Sheet vinyl: Sensation Courtyard in Light Blue Spruce (55020) and Terra Cotta (55021), $26–29 per square yard, installation cost varies
Congoleum Corp.

Carpeting and area rugs: $19.99 per square yard
Shaw Industries Inc., PO Drawer 2128, Dalton, GA 30722; 800-441-7429.

Epoxy tile: Permatop Liquid Binder with Color Quartz in QS8 (red) and QS3 (brown); approx. $5–9 per sq. ft., installed
Permagile Industries Inc., 101 Commercial St., Plainview, NY 11803; 800-645-7546.

FOR MORE INFORMATION

National Kitchen and Bath Association
687 Willow Grove St., Hackettstown, NJ 07840; 908-852-0033.

FURTHER READING

This Old House Kitchens: A Guide to Design and Renovation by Steve Thomas and Philip Langdon, 1992, 288 pp., $24.95
Little, Brown & Co. Inc., 34 Beacon St., Boston, MA 02108; 617-227-0730.

Stainless steel double bowl: Ravinia #K3224, $588.90; pictured with hardwood cutting board #K3280, $58.55; and coated wire basket #K3280, $63.55
Kohler Co., Kohler, WI 53044; 800-456-4537.

Chrome wire basket #L31007
Kallista, 2701 Merced St., San Leandro, CA 94577; 510-895-6400.

Composite 1½ bowl: Asterite #KADC-2233, $424
Kindred Industries, 1000 Kindred Rd., Midland, Ontario L4R 4K9, Canada; 800-465-5586.

Hammered brass single bowl: Hammertone #CS450, $565
Bates & Bates, 3699 Industry Ave., Lakewood, CA 90712; 800-726-7680.

Stainless steel double bowl: Prestige Plus #PRX 660, $765
Franke Inc., Kitchen Systems Division, 212 Church Rd., North Wales, PA 19454; 800-626-5771.

Americast double bowl: #7145, $309-417
American Standard Inc., 605 S. Ellsworth Ave., Salem, OH 44460; 800-524-9797.

Fireclay black medium square: #20293, $290; disposal bowl #L20294, $204
Kallista.

Fireclay single bowl white, with decorated apron: Interlace Alcott #K14571-FC, $1,358
Kohler Co., Kohler, WI 53044; 800-456-4537.

Solid-surface double bowl with drainboard: Swanstone EuroKitchen #KSEU-3020, includes plastic colander, $430
The Swan Corporation, One City Centre, St. Louis, MO 63101; 314-231-8148.

Solid-surface double bowl: WilsonArt Gibraltar #BD323, $525
WilsonArt International, 2400 Wilson Place, Box 6110, Temple, TX 76501; 800-433-3222.

Soapstone single bowl with backsplash: custom-built, $800-1,200
Vermont Soapstone Co., Box 168, Stoughton Pond Rd., Perkinsville, VT 05151; 802-263-5404.

Vikrell double bowl: White Waterstone Workstation #CV3322DBG, $250-300
Sterling Plumbing Group, 2900 Gulf Rd., Rolling Meadows, IL 60008; 800-895-4774.

SINK SAMPLES

1. Composite, one and a half bowls. Resembles enameled cast iron but weighs less. Color-through material means chips and scratches shouldn't show. Lighter colors are more likely to show staining and scorching.
2. Hammered brass, single bowl. Flashy, opulent and expensive, brass is durable but best reserved for sinks that are seldom used. It dents and requires polishing to maintain its luster.
3. Stainless steel, double bowl. Stands up under heavy use and is highly resistant to stains and heat. Choosing 20 gauge or heavier 18 gauge helps prevent dents. Look for an insulating coating on the exterior to dampen noise.
4. Americast double bowl. Porcelain enameled over a proprietary metal and composite base. Americast looks like cast iron, but it's lighter and less costly.
5. Fireclay, square and disposal bowl. The finish is as durable as porcelain enamel, but the sink may crack and chip if something heavy is dropped in it.
6. Fireclay, single bowl. Looks like cast iron but is significantly lighter and normally less expensive. Unlike cast iron, it can be intricately detailed, as in the design on the apron of this pantry sink.
7. Solid-surface, double bowl with drain board. Stains, gouges and scratches are easily repaired by scrubbing with a nylon pad. But it's not as resistant to heat as other materials and may scorch.
8. Solid-surface, double bowl. Seamless joint is possible between solid-surface countertop and sink, leaving no place for dirt to lodge.
9. Soapstone, single bowl with backsplash. Made from stone slabs ¾ to 1¼ inches thick. Holds heat well and is easy to maintain—can be cleaned with just about anything. Heavy (this one weighs 300 pounds) and generally custom made.
10. Vikrell double bowl. Fiberglass makes this composite more resistant to heat and stains than other composites. Available in matte or gloss finish.

1

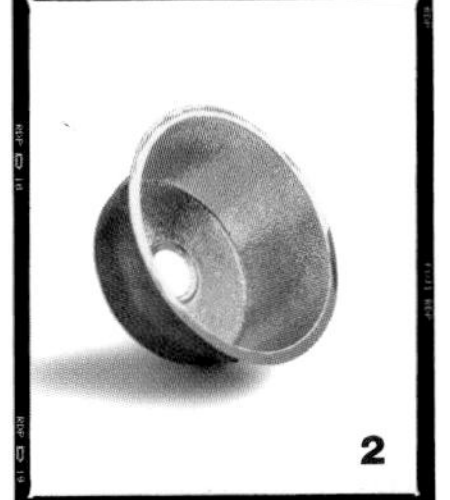
2

3

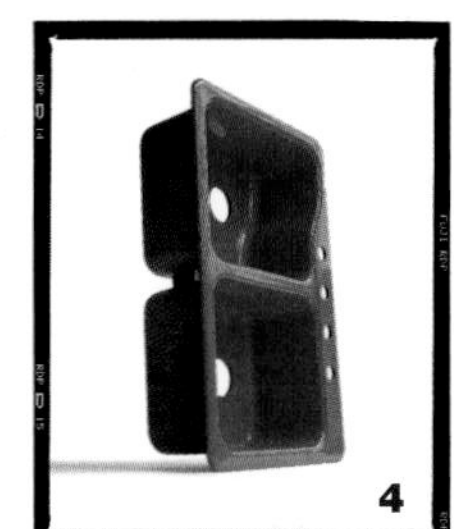
4

5

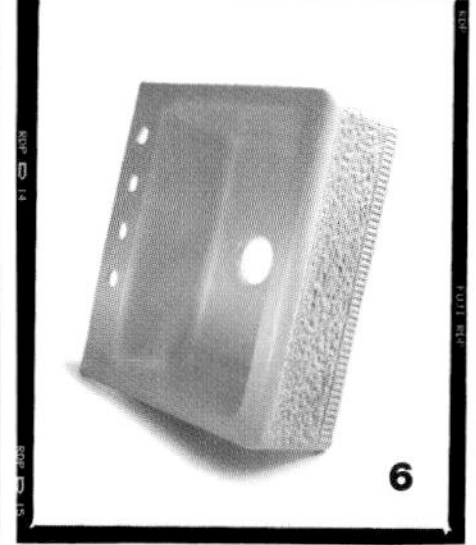
6

7

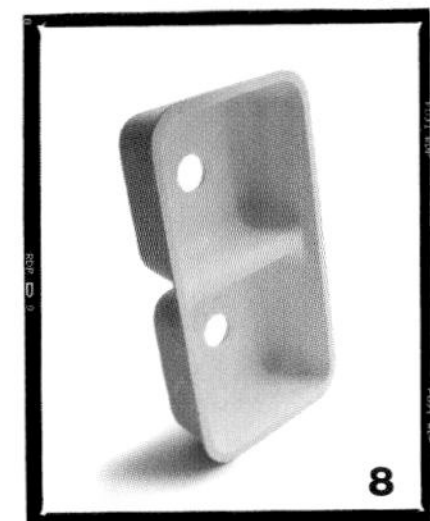
8

9

10

CARE AND CLEANING

One look through a microscope at the gruesome assortment of microbes and bacteria that thrives in the kitchen sink is enough to inspire anyone to keep it well-scrubbed. But what's the best way to clean without ruining the finish? As a rule, surfaces should be wiped daily with a nonabrasive cleanser and a sponge or a soft cloth. Anything scratchier mars the finish and leaves the material beneath susceptible to stains. Solid surfacing and stainless steel with a brushed finish benefit from a light buffing with a nylon mesh pad to remove scratches and, in the case of solid surfacing, stains or scorch marks. Mirror-finish stainless steel can be kept shiny with silver or copper polish. Avoid discoloration in the first place by promptly rinsing away coffee grounds, tea, tomato juice and anything else that stains. There are as many ways to remove stains as there are recipes for salsa. One suggestion: Soak the area with a poultice of vinegar, mild laundry detergent and water.

STEVE SAYS

- Good kitchen design starts with the sink. Place the sink first, and lay out the rest of the work triangle from there.
- Two sinks—one for cleanup, one for prep work—are better than one.
- Bigger is usually better.
- Copper looks great until the first time someone runs water in it. Then it requires polishing and a gentle hand to be sure it doesn't ding.
- Stainless steel complements commercial appliances, holds up well, is affordable and easy to clean.

TYPES OF MOUNTINGS

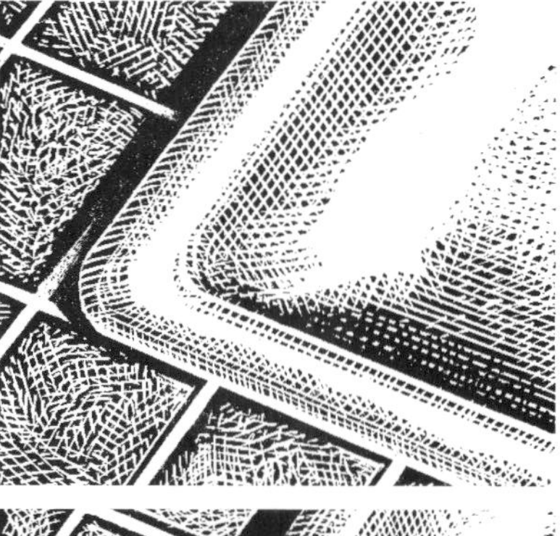

Self-rimming sinks are easy to install: Just plunk them into a hole routed out of the countertop. Caulking between the sink and the counter forms a watertight seal.

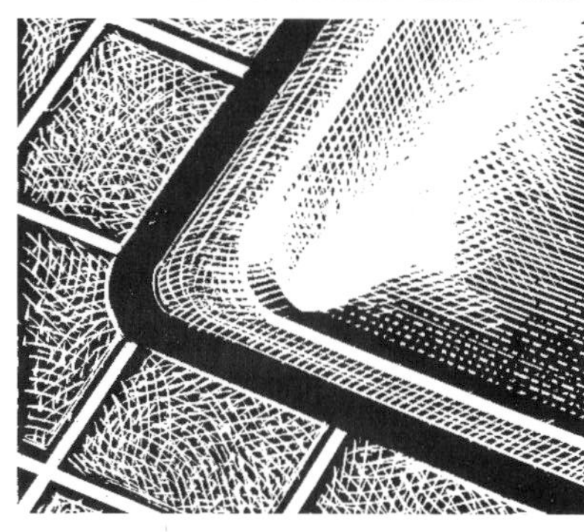

On a flush-mount sink, the lip between the sink and the counter is not as pronounced. Like self-rimming sinks, these are easy to install.

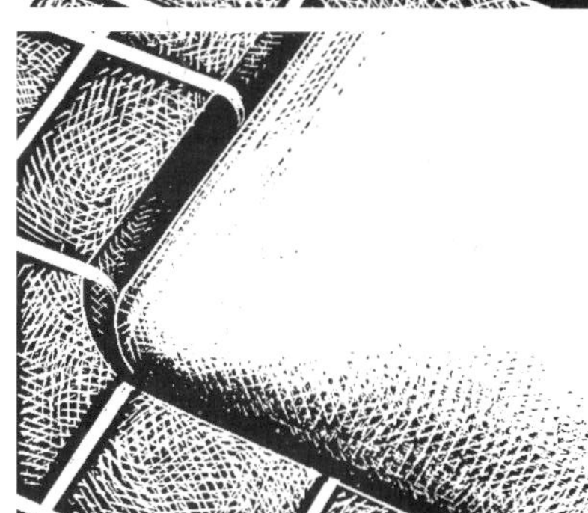

Undermount sinks mean crumbs and spills can be wiped directly from the counter into the sink—there's no lip in the way. Installation is tricky; the hole in the countertop must be carefully cut. Be sure the countertop material is well sealed to prevent it from soaking up water. Undermounts work best with solid surfacing.

showerheads and fittings

With federal water-conservation laws limiting new showerheads to a flow of 2.5 gallons per minute, plumbing expert Richard Trethewey suggests installing a low-flow head as a weekend project. "It sounds like a job you can knock off during commercials on a sports telecast, but there's more to it than screwing the old one off and screwing the new one on," Richard says. "If you're not careful, you'll have more than a weekend project on your hands. And, he adds, "you can update your shower by replacing the old faucet—the kind with separate hot and cold valves—with an antiscald or temperature-control valve. Most homeowners don't realize it can be done without ripping out all their tiling."

If Richard had his way, "there'd be an antiscald valve in every shower in the country." The reason is safety. Flushing a toilet, using the sink or starting the dishwasher can divert cold water from the shower. Without an antiscald valve, you'll get a blast of hot water—up to 140 degrees Fahrenheit, which can cause third-degree burns in a matter of seconds.

Antiscald valves are made for both tub and stall showers, and they're simple to use. Set the desired temperature on the valve, wait a bit for the mechanism to blend the hot and cold flows, then step in and lather up.

SOURCES

SHOWER FITTINGS (HEADS, PERSONAL SHOWERS AND ANTISCALD VALVES)

American Standard Inc.
1 Centennial Ave., Piscataway, NJ 08855-6820; 800-524-9797 or 908-980-3000.

Delta Faucet Corp.
Box 40980, Indianapolis, IN 46280; 800-345-3358 or 317-574-5663.

Grohe America Inc.
241 Covington Dr., Bloomingdale, IL 60108; 708-582-7711.

Interbath Inc.
665 N. Baldwin Park Blvd., City of Industry, CA 91746; 800-828-7943 or 800-423-9485 (in CA).

Kohler
444 Highland Dr., Kohler, WI 53044; 800-456-4537.

Moen Inc.
25300 Al Moen Dr., North Olmsted, OH 44070-8022; 800-553-6636.

Pollenex
217 E. 16th St., Sedalia, MO 65301; 800-767-6020.

Regent International Inc., (Ugenex and Premier brands)
600 Industrial Dr., Cary, IL 60013; 800-210-7054.

Resources Conservation Inc.
Box 71, Greenwich, CT 06836; 800-243-2862 or 203-964-0600.

Symmons Industries Inc.
31 Brooks Dr., Braintree, MA 02184; 800-796-6667.

Teledyne Water Pik
1730 E. Prospect Rd., Fort Collins, CO 80553-0001; 800-525-2774.

Wall-mounted soap dispenser: #71450, $29.99
Better Living Products, 150 Norfinch Dr., Toronto, Ont., Canada M3N 1X9; 800-487-3300.

The Bath Jar: #8324-2, single-mount chrome, $39.99; #8325-2, double-mount chrome, $59.99; #8324-3, single-mount brass, $49.99; double-mount brass, $74.99
Earth Preserv, 500 Decker Dr., Suite 204, Irving, TX 75062; 800-932-7849.

Scald-Safe showerhead
Resources Conservation Inc., Box 71, Greenwich, CT 06836; 800-243-2862 or 203-964-0600.

CHOOSING A SHOWERHEAD

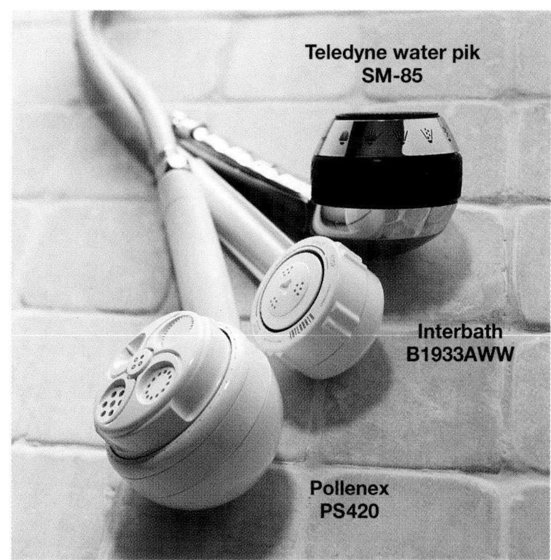

A good guide to low-flow showerheads is the February 1995 Consumer Reports, which rated 29 models. Among regular showerheads, Teledyne Water Pik's SM-62-P ($42) and SM-82-W ($59) finished one and two, ahead of models priced as high as $51 and $65. Resources Conservation's ES-181 and Pollenex's PS50 (both only $7) also beat the $51 head. Among handheld or "personal" models, Teledyne's SM-85 ($69) was first, followed by Pollenex's PS420 ($35) and Interbath's BV923AWW ($59), which tied. Caveats: One, low water pressure can halve the legal 2.5-gallon flow, dampening your spirits but little else. (Solution: a head with a removable flow-restrictor ring.) Two, the risk of scalding increases if a low-flow head is installed but a temperature-control valve is not. That's because the too-hot water backs up behind the head and takes longer to get through.

OTHER SHOWER GADGETS

There is an almost endless variety of shower fittings. Grohe's antiscald Chiara model ($650), for tub showers, offers the simplest installation. No tile-cutting here: Just shut off the water supply, remove knobs and valve stems and screw the fixture onto the hot and cold valve bodies. Another easy option is Resources Conservation's Scald-Safe thermostatic showerhead ($14). It's the least expensive if also the least convenient: It just shuts off too-hot water rather than controlling temperature. Other fixtures also provide safety and comfort. Soap-on-a-rope is one answer to the danger of soap on the floor; a better answer is a wall-mounted dispenser for liquid soap, shampoo and lotions, whether plainly functional, like Better Living's The Dispenser ($25), or decorative, like those from Earth Preserv ($40 to $75). To remove skin-drying chlorine, install a Rainshow'r filter ($40), which screws on between the shower arm and head. For the young, elderly and handicapped, consider grab bars and wall-mounted fold-down shower seats like those from American Standard.

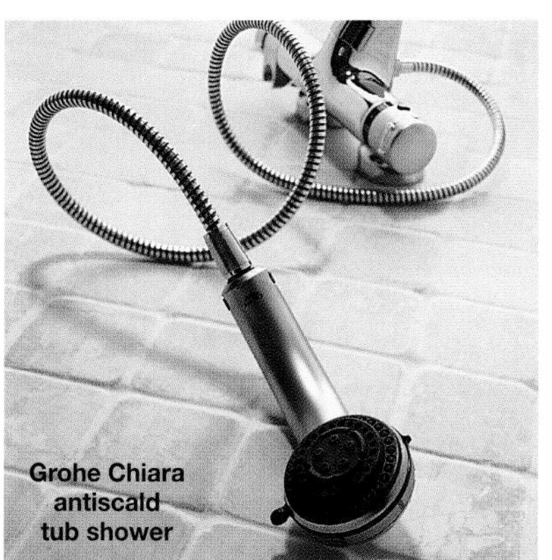
Grohe Chiara antiscald tub shower

Resources Conservation Scald-Safe showerhead

Rainshow'r chlorine removal filter

TOOLS AND MATERIALS FOR INSTALLATION

Don't be daunted by the many tools listed. Several are included just in case: You can't know exactly what you'll need until you can see inside the wall.

Tools: wrenches (adjustable, vise-type, pipe, channel-type); variable-speed power drill and ¼-inch masonry bit; hammer and sharp cold chisel; electric tile cutter (optional); propane torch with fine or "pencil" tip; screwdrivers (Phillips and plain); fireproof plumber's cloth; handle puller (for removing shower knobs); tubing cutter (a small "imp" for working in tight spaces).

Materials: Teflon pipe tape; lead-free solder; flux; bucket of water and/or small fire extinguisher; duct tape; silicone caulking; emery cloth.

Equipment: antiscald valve; modification plate; low-flow head; a foot or two of ½-inch copper tubing (Type L) and half a dozen miscellaneous copper fittings (elbows, couplings, 45-degree angles, copper unions or slip couplings; assorted "ream-style" adapters in the event that you have brass pipes).

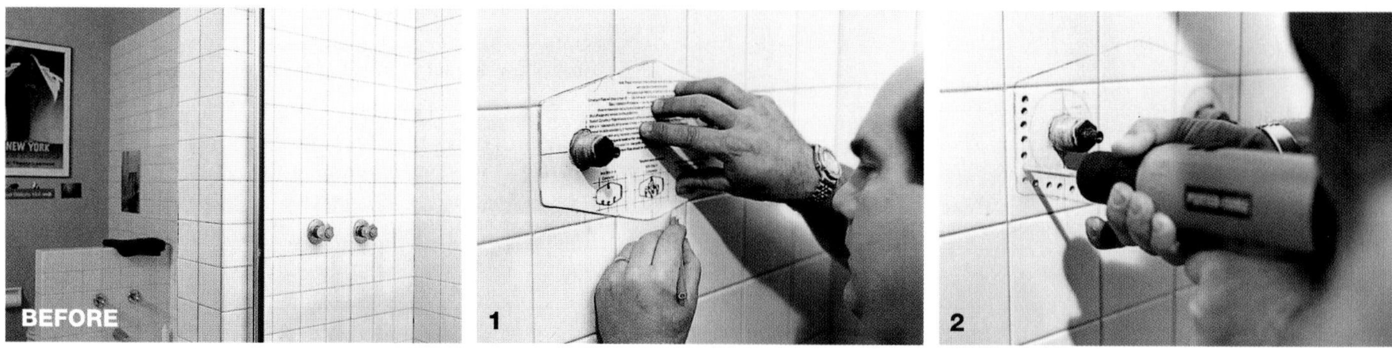

First shut off the hot and cold water lines to your bathroom and open all bathroom taps to break the vacuum in the pipes. Then pry the index caps (marked "hot" and "cold") from both knobs, remove the screws underneath and take off the knobs. If the knob simply slides off (unlikely), fine. If not, use the handle puller. Unscrew the escutcheons (decorative sleeves covering the valve bodies) and flanges (trim rings behind the escutcheons). **1.** Center the modification plate's paper template: Position it over the valve stems, press down hard with your fingers and poke a hole through it into the stems' screw holes. On a countertop, center an escutcheon over each hole, outline it in pencil and cut out the circle. Position the template over the valves and outline it with pencil or china marker. The outline marks the no-go area: Don't go outside the lines. **2.** Drill closely spaced pilot holes through the tiles around the valves. To start, press the bit firmly against the tile and drill slowly so it won't skip erratically across the surface.

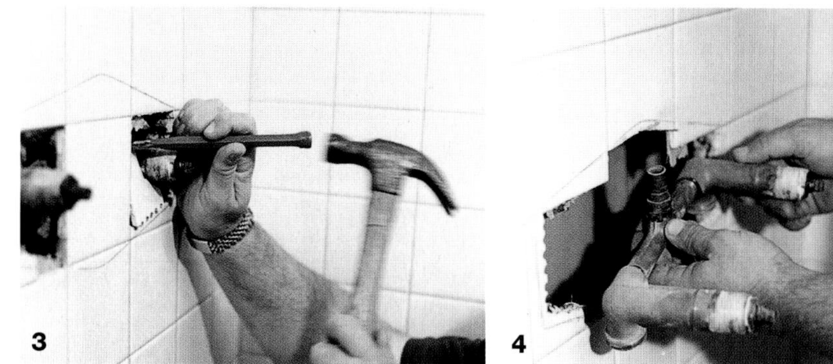

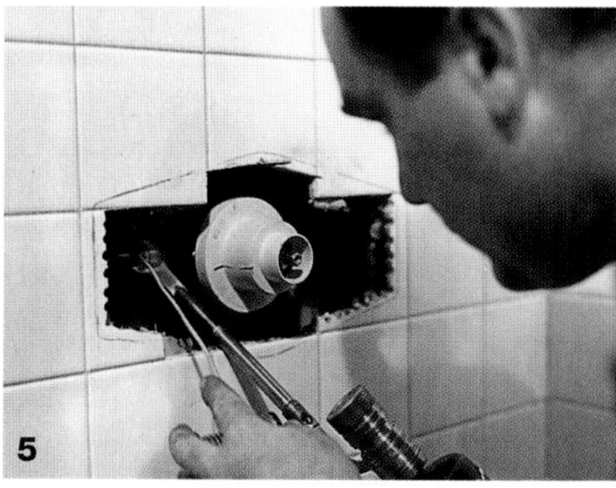

3. Chisel out the tile: a few blows on one side of each hole, a few on the other. Or use an electric tile saw to cut within the outline. **4.** Now you can see the three joints that must be disconnected. Rising from the middle of the faucet body is the supply tube for the showerhead. Cut it with an imp cutter (or a hacksaw blade—but watch your knuckles). At left and right are the hot and cold water supplies. If they're sweat-soldered, a torch will unsweat them. If they're "union" joints (with large locknuts), use a wrench. Remove the faucet body and, if necessary, unsweat the rest of the union fitting on each supply (use pliers to handle these hot parts). When using a torch, it's wise to put a fireproof plumber's cloth behind the work. **5.** Install the antiscald valve. Reconnect the hot, cold and shower piping to the new valve body, using appropriate copper fittings and cutting tubing as necessary. Often it's difficult to make these connections in such a confined space. Two special fittings may be required: a copper union or a copper slip coupling. Clean all tubing and fittings with emery cloth and apply flux at each joint. Use the plastic rough-in plate (which resembles a cup) to make sure the valve is level and plumb. The "service stops" (screwdriver-slot shutoffs on the front of the valve) should be fully open (counterclockwise). The spindle (central valve stem) should be removed. Apply torch to each fitting until the flux bubbles, then apply solder to the opposite side until the fitting is filled. You'll know it's filled when you see the solder come toward the front or flame side.

6. Flush the valve to remove any flux, solder or sediment. The following steps may seem obscure and confusing, but trust Richard and do exactly as he says. **(a)** Close the service stops and turn on main hot and cold supplies. **(b)** Slowly open the service stops, allowing water to flush the valve. **(c)** Close the stops, reinstall the spindle and reopen the stops. If all is well—no leaks at the joints—install the modification plate and trim. Apply silicone caulking or plumber's putty to the back of the plate. The plate comes with two bolts and two spanner bars which should be tightened equally (don't overtighten). Follow manufacturer's instructions for installation of trim ring, escutcheons and handle. **7.** Remove the old showerhead from the pipe ("shower arm"), which is threaded at both ends. Over time, threads can fuse; if you force them, they'll break inside the wall, ruining both your weekend and your budget. Avoid this by using two wrenches: one on the shower arm, the other on the head. Unless you really want the neighbors to know you did the job yourself, use duct tape so the wrenches won't scar the chrome. Holding the shower arm steady with one wrench, gently unscrew the head with the other. Screw on the new head, taking the same precautions. If it lacks a washer, wrap the threads with a turn or two of Teflon pipe tape to prevent drips. Personal showers install the same way, but some have a wall fitting for the head. Using the fitting as a guide, mark screw holes, drill them, insert anchors (included with the shower) and mount the fitting. Test for leaks; remove duct tape and excess caulking. Put your tools away and take a shower. You'll need one about now. Note: For fiberglass showers, the steps are largely the same. Make shallow cuts to avoid pipes. For maximum control, use a jigsaw with a fine-tooth blade.

Rainshow'r chlorine filter #RS-502, $39.95; #RS-502DS, showerhead and filter, $59.95
Pacific Environmental Marketing & Development Co., 421 S. California St., Unit D, San Gabriel, CA 91776; 800-243-8775.

TOILETS, FITTINGS AND PARTS

American Standard Inc.
1 Centennial Ave., Piscataway, NJ 08855-6820; 800-524-9797 or 908-980-3000.

Gerber Plumbing Fixtures
4600 W. Touhy Ave., Chicago, IL 60646; 708-675-6570.

Hunter Plumbing Products
1775 La Costa Meadows Dr., San Marcos, CA 92069; 800-486-8371.

Kohler
444 Highland Dr., Kohler, WI 53044; 800-456-4537.

Sloan Valve
10500 Seymour Ave., Franklin Park, IL 60131-1259; 708-671-4300.

Toto Kiki
415 W. Taft Ave., Orange, CA 92665; 800-877-1541.

Clivus Multrum
Clivus Multrum Inc., 104 Mt. Auburn St., Cambridge, MA 02138; 800-425-4887.

Incinolet
Research Products/ Blankenship, 2639 Andjon Dr., Dallas, TX 75220; 800-527-5551.

Microflush
Microphor, 452 East Hill Rd., Box 1460, Willits, CA 95490; 800-642-7674.

Sun-Mar
5035 N. Service Rd., Unit C-2, C9-10, Burlington, Ontario, Canada L7L 5V2; 800-461-2461.

Ultra-Flush
35 Citron Ct., Concord, Ontario, Canada L4K 2S7; 905-738-0055.

FOR MORE INFORMATION

"Low-Flow Toilets," Article #9994, $7.75
Consumer Reports; 800-766-9988.

low-flow toilets

Taken for granted, covered with euphemisms and hideous fuzzy tops, toilets aren't usually the focus of attention. But that's changing now that federal law requires new models to use only 1.6 gallons per flush—a marked decrease from the 3.5 gallons or more currently going down the drain.

The big decision about low-flow toilets, manufactured since 1994, isn't which style to buy (The Patriot? The Pillow Talk? The Trocadero?) or the designer color (Fawn Beige? Tender Gray? Innocent Blush?) Essentially, there are three types: gravity-operated (the vast majority), pressure-tank models (increasingly popular) and the rest (obscure, often very expensive).

New gravity models use refined fill mechanisms to limit inflow and improved flush-valve, bowl and drain designs to meet the 1.6-gallon requirement. Some contain 3.5 gallons but use the excess only to develop "head pressure," which powers the flush; shutoff valves close after the legal limit.

The pressure-tank toilet, commonly (but incorrectly) called a "flushometer," looks about the same as the gravity type, but inside its standard ceramic tank is a sealed pressure tank. During filling, water enters it under line pressure, compressing the air within. During flushing, the compressed air expands, rapidly and violently forcing water into the bowl and down the drain, accompanied by a sonic boom that could loosen your fillings.

But noise is relative; listen before you write off pressure tanks. Most makes sound alike because almost all manufacturers install the same Sloan Flushmate pressure tanks. Kohler's proprietary tank is notably quieter.

Where water pressure is low (below 20 to 25 psi static), go with gravity-operated toilets. Where house-to-sewer drains are long (50 feet or more) or prone to clogging, the extra oomph of a pressure tank can help.

Many manufacturers offer only one flush system. American Standard and Crane are among the big names that offer both, and Kohler offers two varieties of each. A refinement of Kohler's basic gravity model has a cam-actuated flush valve that lets you choose a standard 1.6-gallon flush or a 1.1-gallon flush (for liquids only). Most Kohler pressure-assist units use air tanks, but the top-of-the-line models have submerged electric pumps. They're expensive, but very quiet.

When it comes to choosing an individual model, consumers face two problems. One is that water-miser toilets have no "buy points"—that is, specific features that make them must-haves or must-avoids. There are things to consider, though, such as the size of the "pond," or water patch in the bowl. Larger is better. Small ponds mean frequent cleaning. And check the height of the seat on the manufacturer's spec sheet: some are higher than the standard 14 inches.

The other problem is that price is no guide to performance, so get the February 1995 Consumer Reports "Super Bowl" issue, which rated 32 models from 11 makers. It found that Gerber's $210 Ultra Flush 21-302 performed just as well as an $815 famous-name competitor. Among other surprises, one big-name manufacturer had three models flunk for excess consumption.

"Function comes first and fancy comes second," advises *This Old House* plumber Richard Trethewey. "Plumbers sometimes get bad-rapped as nonprogressive for not being interested in color and styling, but those aren't primary concerns. Get the toilet that does the best job."

Unfortunately, saving water costs money, so be on the alert for local and state rebate programs. Some pay the whole cost of replacement, depending on the make and model selected. Those in New York City and Los Angeles have been fairly well publicized, but even so, many people don't realize that single-family houses qualify for the rebates, not only apartment buildings and offices. Finding out about rebate programs takes persistence; you may have to call regional water boards to cover all bases. A good informed source would be your plumbing supplier, as rebate programs represent revenue opportunities he's unlikely to overlook.

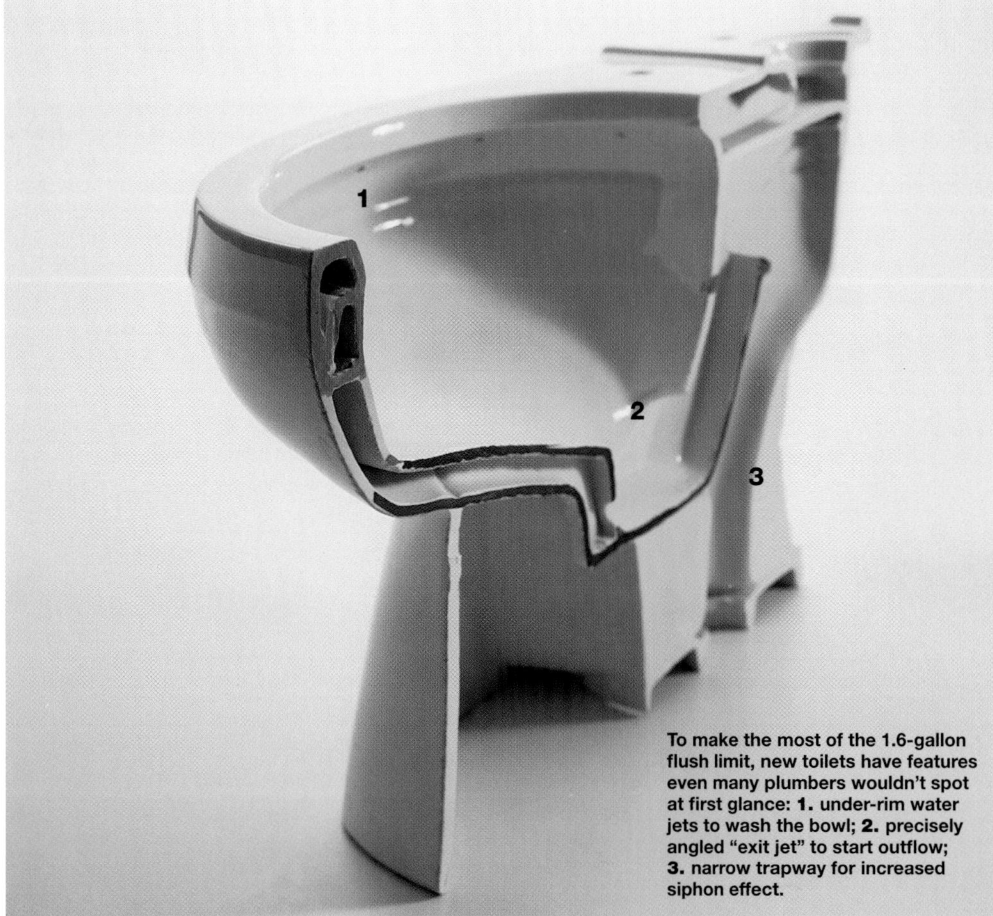

To make the most of the 1.6-gallon flush limit, new toilets have features even many plumbers wouldn't spot at first glance: **1.** under-rim water jets to wash the bowl; **2.** precisely angled "exit jet" to start outflow; **3.** narrow trapway for increased siphon effect.

SOURCES

Installing & Repairing Plumbing Fixtures, **by Peter Hemp, 1994, 184 pp., $19.95**
The Taunton Press, 63 S. Main St., Box 5506, Newtown, CT 06470-5506; 800-888-8286.

The Straight Poop, **by Peter Hemp, 1986, 176 pp., $11.95**
Ten Speed Press, Box 7123, Berkeley, CA 94707; 800-841-2665.

The Toilet Papers, **by Sim Van der Ryn, 1995, 127 pp., $10.95**
Ecological Design Press, Ecological Design Institute, 10 Libertyship Way, Suite 185, Sausalito, CA 94965; 415-332-5806.

Basic Plumbing, **1995, 96 pp., $9.99**
Sunset Books, 81 Willow Rd., Menlo Park, CA 94025; 800-634-3095.

Replacement parts for low-flow toilets cost more too, because most are specific to make and model. Hunter Plumbing, Fluidmaster and other major manufacturers produce competitively priced replaceables such as fill mechanisms and flush valves (also called flappers). The risk for the homeowner is in using cheap generic parts found in hardware-store blister packs. These will malfunction unless designed for the toilet they're used in.

Since even the best miser toilets don't always flush as well as the water-wasters, take a hint from the experts: Use less toilet paper (a major cause of clogs). Options that eliminate paper (while improving hygiene and assisting the elderly and handicapped) are Toto's Washlet SIII and Zöe toilet-seat bidets, which use electrically operated water sprays. Another trouble-saving choice is Kohler's Peacemaker, a seat that flushes the toilet electrically (and only) when closed.

What about other toilets? They're mostly for extreme conditions. Sim Van der Ryn, author of *The Toilet Papers*, recommends composting toilets like the Sun-Mar and the euphoniously named Clivus Multrum, which return the soil to the soil, so to speak, without water. The Incinolet, which burns waste, requires electricity, as does the Microflush, whose air compressor provides a quiet half-gallon flush. Simplest is Canada's economical Ultra-Flush, adapted from motor-home designs, which uses only one quart of water. Those with difficult septic tanks and backwoods cabins, take note.

SAVING WITHOUT REPLACING

You don't have to buy a new toilet to save water. Seepage through the flush valve can be detected by putting food coloring in the tank. Don't flush; if color later appears in the bowl, you're wasting approximately 30 gallons a day. If the seepage is actually loud enough to hear, the loss rate is more like 250 gallons a day.

In either case, the fix can be as easy as scrubbing the metal seat of the flush valve with steel wool to remove accumulated gunk. Or buying a new flapper valve or flush ball, which can be installed in a matter of minutes.

Kohler's selectable-flush mechanism, which allows you to choose a regular flush or 1-gallon flush, has been imitated by parts manufacturers. The generic versions save about half a gallon with each "short flush." But note that one size does not fit all, and neither does one shape. Remove and measure your current flush handle before buying, bearing in mind that it probably has a left-hand (counterclockwise) thread.

Other modest savers include water dams, which isolate part of the tank, and bricks. Dams can be finicky to fit, and bricks can leach gunk into the tank; instead, use a half-gallon jug of water, with some stones for ballast. If the jug interferes with your tank's fill mechanism, don't be tempted to make room by manhandling the ballcock or other parts.

"That stuff has been underwater for years," Richard says, "and that makes the parts brittle."

What about reducing inflow by fiddling the fill? Bad idea. That reduces the water level and thus the head pressure necessary for an effective flush. Jugs work by displacing water to reduce inflow without affecting head pressure.

Can owners of Victorian houses save water and maintain historical correctness by installing Sloan Flushmate pressure-assist units inside the wall tanks of their toilets? In a word, no. Those tanks, mounted overhead, develop abundant power already; adding pressure-assist will produce the dread "Geyser Effect."

REPLACING A TOILET

STEP 1

STEP 5

STEP 9

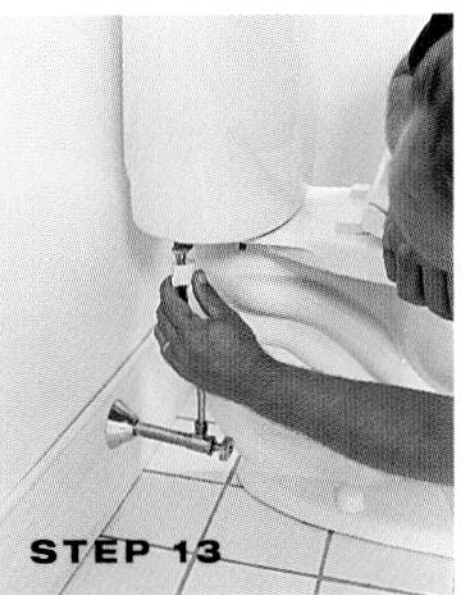
STEP 13

STEP 16

What Richard does in an hour will take you a little longer. For efficiency's sake, read through the process twice and start work early (but never on a holiday; you want the plumbing-supply store open).

Removal: 1. Shut off the angle stop (the small water valve protruding from the wall below the left side of the tank). 2. Flush the toilet and sponge out remaining water from the tank. 3. Remove the supply tube leading from the angle stop to the tank. 4. If your toilet is a one-piece type, proceed to step five. If it's "close-coupled"—tank bolted to bowl—unscrew the bolts and lift off the tank. Some tanks are screwed to the wall and connected to the bowl by a flush pipe. Unscrew this pipe, "or simply slice it with a hacksaw—that's the quick-and-dirty way," Richard says. "My favorite." Now sit on the bowl backward to remove the tank, letting it settle gently onto your knees as the last screw comes out. 5. Remove caps from closet bolts and unscrew the nuts. If they're corroded, use a hacksaw or penetrating oil. 6. Tilt the bowl forward to avoid spilling residual water, then remove. Stuff a large wad of newspaper into the soil pipe to block sewer odors. 7. Inspect flooring around the closet flange (where the toilet joins the soil pipe) for rot caused by seepage. "That," says Richard, "would be your cue to grab a Bud and call your plumber."

Installation: 8. Replace the old wax closet seal and closet bolts with new parts. (Richard prefers wax seals; they're cheap, and one size fits all. Rubber seals cost more, though they are more "beginner-tolerant.") 9. Gently set the new bowl in place. Sit on it, compressing the seal for a tight fit, then tighten the nuts with your fingers. Continue pressing the bowl down and tightening the nuts alternately. As resistance increases, level the bowl, shimming if necessary. Overtightening nuts can crack the fixture, so tighten nuts further during the next few days, as the toilet settles. 10. Pour buckets of water into the bowl to check for leakage. 11. Now mount the tank: Press the large gasket over the spud projecting from the tank and set the tank onto the bowl. Line up the bolt holes and insert bolts and washers, tightening the nuts from underneath. Tighten them alternately, using a screwdriver and wrench. Don't overtighten. 12. Fasten chain to flush handle, leaving minimal slack. 13. Reconnect the supply tube: If the new tank is higher or lower than the old, the original supply won't fit. For amateurs, a flexible supply usually solves the problem; tighten nuts gently at angle stop and fill mechanism until supply is firmly seated. 14. Open the angle stop fully. When the tank has filled completely, flush the toilet several times to check that it operates properly. 15. Install the toilet seat. 16. Step back and admire your work.

TOOLS AND MATERIALS

Assemble tools and materials the day before. You will need:

- 8-inch adjustable wrench
- 8-inch straight-blade screwdriver
- spud wrench or 10-inch water-pump pliers (Channellocks)
- hacksaw
- 24-inch level
- solid brass closet and tank bolts (if included with toilet, test with a magnet; if they're just plated steel, replace with brass)
- wax or rubber closet seal
- flexible supply tube
- penetrating oil
- sponges and rags
- old newspapers
- hand cleaner

RICHARD SAYS

Start by measuring the rough-in, the distance in inches from the wall—not the baseboard—to the center of the soil pipe. It's usually 12 inches but sometimes 10 or 14. If you don't know it, one of two things will happen when you buy your new toilet. Either the clerk will ask for it and you'll have to say, "Duh"—or he won't, and he'll sell you a toilet that won't fit.

Toilets are fastened with closet bolts, so measure from the wall to the center of the bolt on one side of the pedestal. Older fixtures have two bolts per side; measure to the nearer bolt. Write the rough-in down in a safe place. Your forehead, for example.

tiling a bathroom

CHOOSING TILE

Tiles are produced from fired clay or cut from natural stone. Most tiles sold in the United States are ceramic tiles, made from clay, ground shale or gypsum and other ingredients such as talc, sand or vermiculite. Ceramic tiles are categorized according to permeability (water absorption) by the American National Standards Institute, which determines how the tile can be used. Categories run from nonvitreous (readily absorbs water) to semivitreous, vitreous and impervious.

In terms of design, tiles are either field tiles (those set in the main field of an installation) or trim tiles (those shaped to border and complete the main field, such as a bullnose). Tiles with hand-painted designs or raised relief shapes are called decoratives. Most tile makers offer a range of trim tiles designed to be used with their field tiles. It is almost impossible to get a trim tile from one manufacturer to match a field tile from another; it's better to work out solutions with what's available in one line. When ordering, bring a measured drawing of the areas to be tiled so a dealer can help design the installation, ensuring that the trim tiles cover the corners and meet the edges properly. The dealer will estimate the number of tiles needed, adding at least 5 percent to provide extras in case of miscuts or breakage and for later repairs.

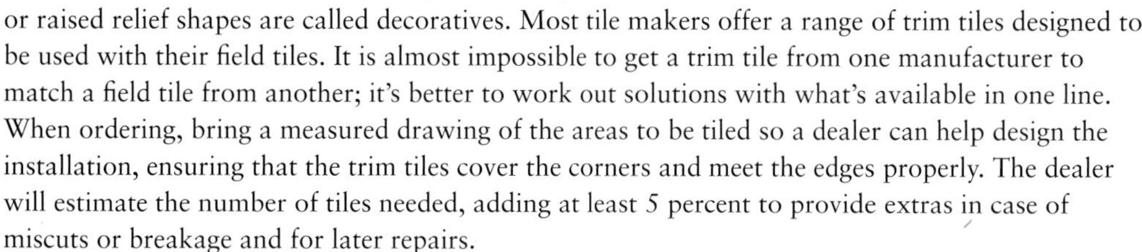

All tile must be laid with consistent spacing. Some tiles come with self-spacing lugs on the edges, determining the width of the joint. If tiles are not self-spacing, try laying them with spacers, small plastic devices in varying widths (from $\frac{1}{16}$ to $\frac{1}{2}$ inch). Spacers are fairly new; tile setters traditionally use anything from nails to a stringcourse.

SIZING AND CUTTING

1. For straight cuts and trimming narrow slices of tile, a diamond-blade wet saw is the surest and easiest tool to use. Too large and expensive for the homeowner's arsenal, it can be rented from most tile dealers or home centers. The diamond-tipped blade is cooled by a stream of water from a recirculating pump. **2.** Tom Silva uses nippers to make a curved cut to go around a pipe fitting; he first marks the cut with a grease pencil, then nips out small bits of the tile until the opening is clean. The hole will be covered by the faucet trim, so a perfect fit is not critical. **3.** For cut tile that will be visible, such as the piece that curves around the tub, nippers can be used to cut the curve, and the untoothed blade of the wet saw can follow up to create a smooth, slightly rounded finish. **4.** Norm tried out a new cordless handheld diamond-blade wet saw that proved quite versatile, combining the clean cuts of the stationary diamond-blade saw with the nippers' maneuverability. "It's great for complex cuts that a nipper couldn't do," he says, "and for doing a few tiles at a time, like a repair job, it beats setting up the big saw." Here Norm cuts a tile freehand.

1

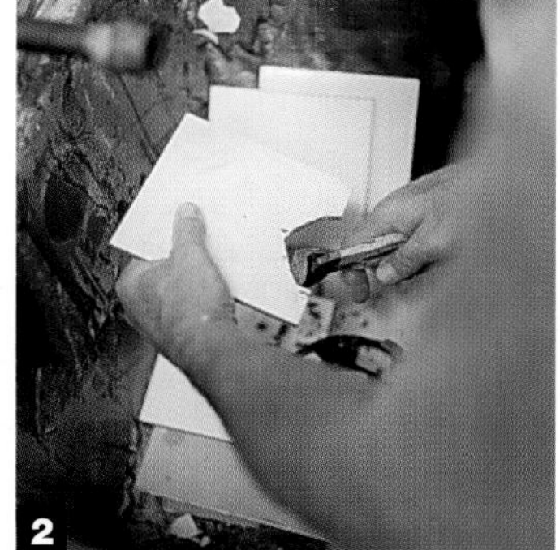

2

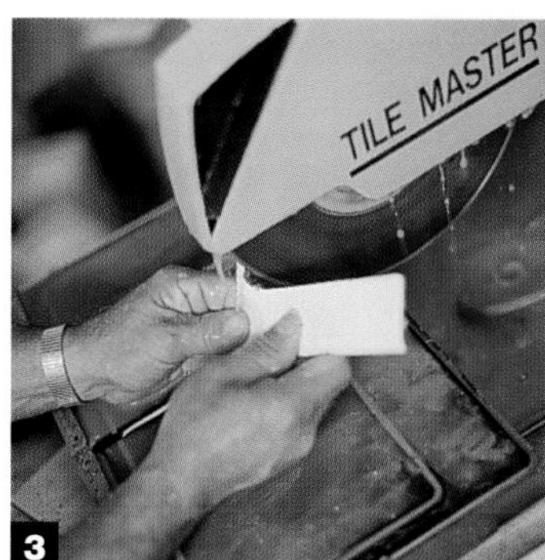

3

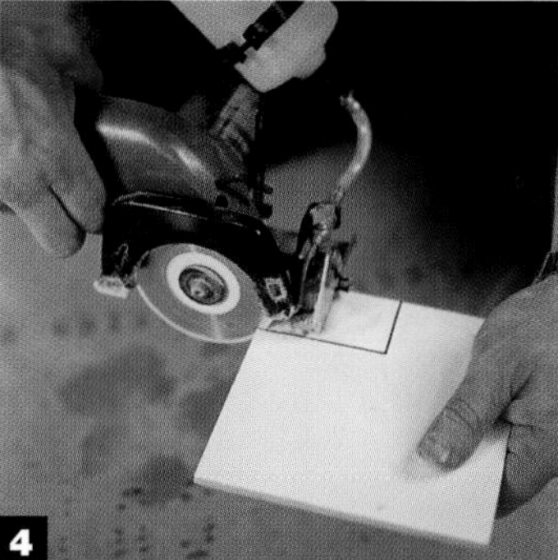

4

SOURCES

Tub sealer: SP-01 Scratch Protection, brushable or sprayable protective coating, $35 per gal.
Surface Protective Products Int'l. Inc., 1205 Karl Court, Suite 116, Wauconda, IL 60084; 800-789-6633.

Decorative Tiles:

MX white 6"x6" glazed wall tile decoratives; seashore hand-painted wall tiles from Spain, $5.25 per tile; sea surf border on white, 3"x6", and sailboats on white, 6x6" wall tile, $20.20 per tile; HP matte white 5¾"x 5¾" field tile, $8.82
Country Floors Inc., 15 E. 16th St., New York, NY 10003; 212-627-8300.

Hydroment multipurpose acrylic latex admixture, #425, 1 gal., $17.58
Bostik, 211 Boston Street, Middleton, MA 01949; 800-726-7845.

Non-sanded, mildew-resistant, white dry tile grout, #WDG5, approx. $4.50 for 5-lb. mix
Custom Building Products, 13001 Seal Beach Blvd., Seal Beach, CA 90740; 800-272-8786.

Tile spacers: ⅛" #LG and 1/16" #LG, around $4 for bag of 300
Walton Tool Co. Inc., 650 W. 16th Street, Long Beach, CA 90813; 800-421-7562.

Silicone caulk: construction tripolymer sealant
Geocel Corporation, Box 398, Elkhart, IN 46515; 800-348-7615.

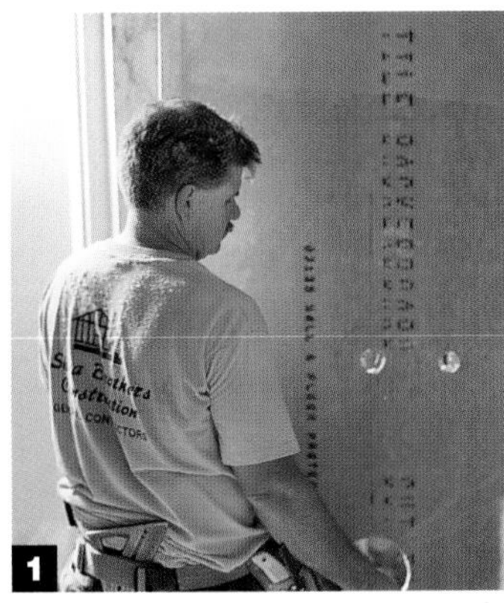

1. Tom Silva lines the framed tub area with cement backer board, a strong substrate designed specifically for tile installation, to a height of 5 feet, well above the wet area. The board has an interior core of sand and portland cement and an outer reinforcing skin of fiberglass mesh. It can be scored and snapped, or cut with a diamond blade if a smooth edge is desired. **2.** He attaches the board using self-drilling screws. The blue coating on the tub is a temporary sealer to protect it from damage during construction. **3.** On the upper wall, which will not be tiled, Tom puts up greenboard, a type of water-resistant drywall. He applies a water-resistant joint compound to the seams with a 6-inch knife. The compound takes 90 minutes to harden. **4.** With a flat trowel, he applies a portland cement mix over the seams and corners, which have been taped with fiberglass tape.

5. Norm Abram begins the tiling by establishing a level line and a plumb line—the key to laying out tile. He sets the first tile at the point where the lines meet. The process is a lot easier in a new area built for tiling than in an existing, out of square area. **6.** Steve Thomas applies enough premixed acrylic adhesive to cover a section he can tile in 10 minutes. He uses the notched side of his spreading trowel, held at a 30-degree angle, to make a ridged base into which he and Norm press the tiles. The depth of the trowel notches should be about two-thirds the thickness of the tiles. **7.** The 6-inch tiles fit snugly together and are self-spacing, although they are slightly irregular and must be worked in place to keep the lines straight. Shims under the starting line of tiles provide space for caulking. **8.** Norm sponges adhesive from the tiles before it cures; if the adhesive dries, thinner is needed to remove it.

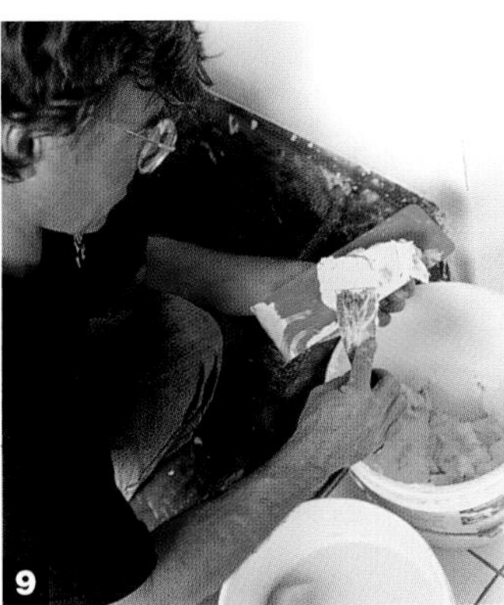

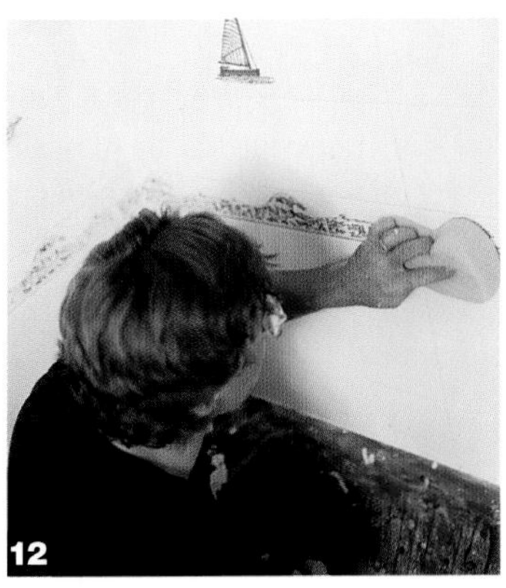

9. Steve prepares a trowel full of mixed grout, then applies it to a small area of the wall. Sand is often used as a filler in grout, but because the joints between these tiles are less than 1⁄16 of an inch, unsanded grout is used to allow a tighter fit. White grout was chosen to blend in with the tiles. **10.** Steve spreads the grout with a float held at a 30-degree angle, making one pass vertically, then another horizontally. **11.** He presses the grout into the tile joints until they are filled. Then he makes a final pass, holding the float diagonally and at a 45-degree angle to the tiles. **12.** After the entire enclosure is grouted, Steve uses a barely damp sponge to wipe down the tiles before the grout hardens. He damp-wipes in a circular motion, taking care not to gouge out the hardening grout. In 24 hours, when the grout is almost dry, he will remove any residue by buffing with a dry cloth.

notes

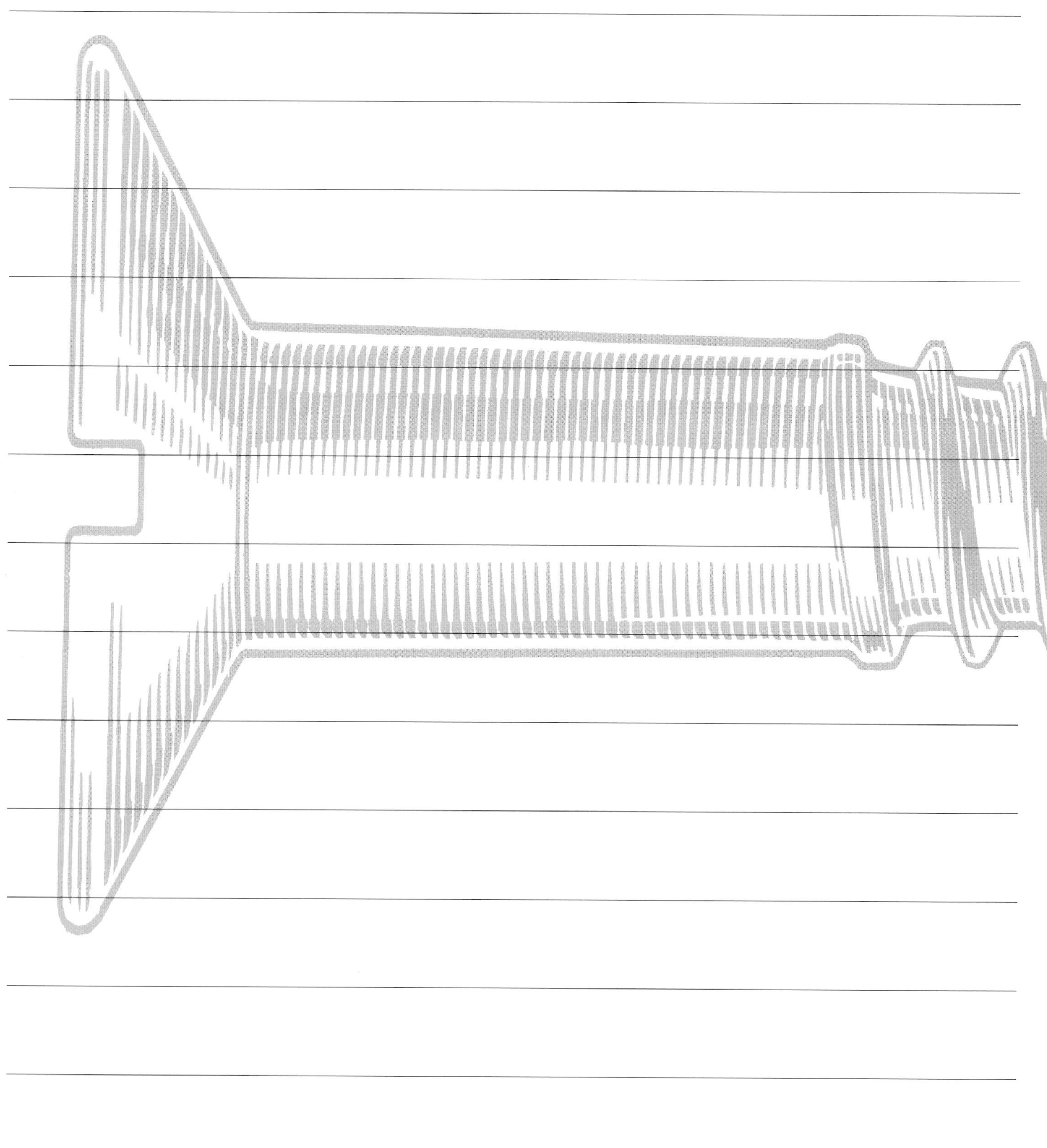

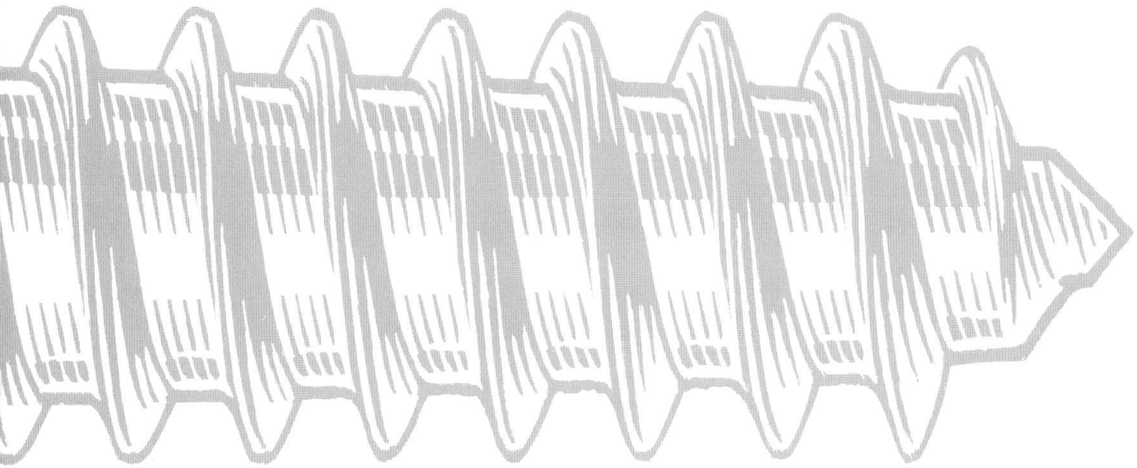

fasteners

NAILS | ADHESIVES | SCREWS | ANCHORS

nails

MATERIALS

Indoors, where rust is rarely a problem, nails made with mild steel do the job. (They are called bright or unfinished nails.) But outside—on porches, roofs and siding—water and air will eventually burn steel into dust. The result might be as benign as unsightly streaks on exterior trim or as catastrophic as the unexpected collapse of a deck. To combat corrosion, consumers have two choices: zinc-coated steel nails or those made with metals that oxidize more slowly. Galvanized nails stave off rust because zinc, a more reactive metal than steel, sacrifices itself to the elements first. But not all galvanizing lasts.

In 1973, the U.S. Forest Products Laboratory in Madison, Wisconsin, pounded nails into pressure-treated boards and subjected them to jungle-moist humidity for 14 years. Hot-dipped was the only type of galvanized nail to survive; electrogalvanized and mechanically galvanized nails nearly disintegrated. The lab recommends that a galvanized nail have at least a 1.7-mils-thick zinc coating. (Useful information, but manufacturers don't print it on the box.) Not even the thickest galvanizing wards off trouble indefinitely. When a zinc coating isn't enough, good contractors turn to:

- Stainless steel, an alloy of chrome, nickel and steel. Stainless lost less than 0.1 percent of its weight in the Forest Products Lab test. It comes in types 304 and 316. The latter is more expensive, more durable and recommended for seaside building. Both types react minimally to redwood, cedar and acid rain. They do rust, but only superficially. The biggest drawback is cost: A 25-pound box of 5d stainless siding nails is $150. The same box of galvanized costs $50.
- Silicon bronze, an alloy including copper and tin. These performed almost as well as stainless, with a 0.6 percent weight loss. Look for them in marine supply stores.
- Aluminum. It's an inexpensive alternative to stainless steel for use on siding, fencing and gutters, but it bends with little provocation and oxidizes quickly in salt air. Aluminum nails dissolve in pressure-treated wood.
- Copper. Too soft to use where strength is important, but perfect on long-lived slate or tile roofs.
- Monel, a nickel-copper alloy. This offers perhaps the ultimate in corrosion resistance—and price. A 25-pound box of 8d ring-shanks costs a whopping $400.

SHANKS

Smooth: Made from plain round wire. The shank of all trades and inexpensive too. (Textured shanks require additional manufacturing steps.)

Barbed: Seldom seen, the result of an early process intended to increase holding power.

Spiral (or drive screw): Shank rotates into wood as it is hammered. Takes more effort to drive but provides excellent resistance to popping. Used for decks, siding and flooring.

Half spiral: Easier to drive but still holds well enough to keep planks flat and squeak-free. A favorite of flooring contractors.

Double spiral: A brute to drive—and remove—it provides maximum holding power. Top-quality spiral nails are cut with a die, not twisted.

Annular (or ring-shank) and **angular:** Individual ribs or rings on the shank engage wood fibers tightly. Often and wrongly called threaded nails, they provide excellent resistance to popping under high loads. But once the load threshold is crossed, popping is sudden. (Smooth shanks surrender more slowly but at lower loads.) Ring-shank nails lose some holding power when hot-dip galvanized because the zinc tends to clog the rings.

Fluted: Straight or slightly spiraled grooves ease penetration of hardened nails into concrete or masonry.

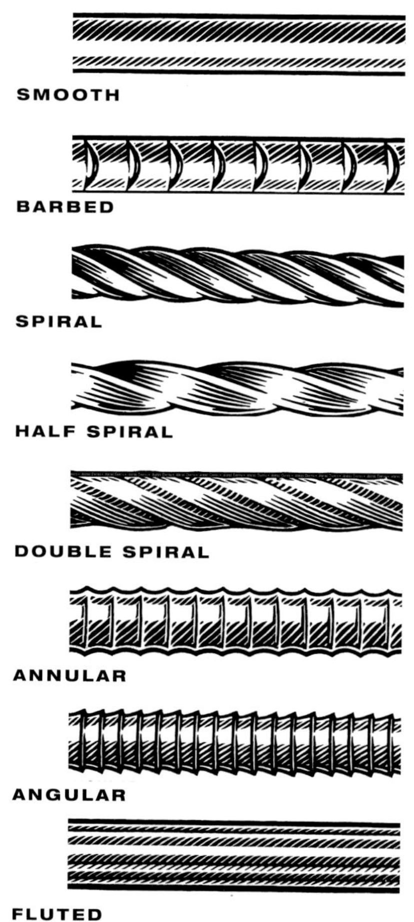

POINTS

Diamond: The most common point style, it eases a nail's entry into wood and wedges fibers apart. The point is formed when nail-making machinery cuts the nail off on four sides, producing the distinctive facets. Cuts from only two sides produce chisel points, found in heavyweight hardware such as railroad spikes.
Long diamond: An extra-sharp point permits easy penetration of drywall. A variant is the needle or needlepoint, which is equally sharp but unfaceted, like the tip of a freshly sharpened pencil.
Blunt diamond: Reduces the risk of splitting wood, particularly when driven through thin stock or near the ends and edges of planks. Blunting regular diamond points by whacking them with a hammer achieves the same result.
Blunt: A squared tip helps prevent splitting by shearing wood fibers rather than wedging them apart. Found mostly on flat-sided cut nails. Blunt-tip round nails are still available from some manufacturers by special order.

DIAMOND

LONG DIAMOND

BLUNT DIAMOND

BLUNT

HEADS

PLAIN
The common head of the common nail, flat on top and bottom.

DUPLEX
Two plain heads on one shank. Upper head remains above the surface to facilitate removal from temporary structures such as scaffolds.

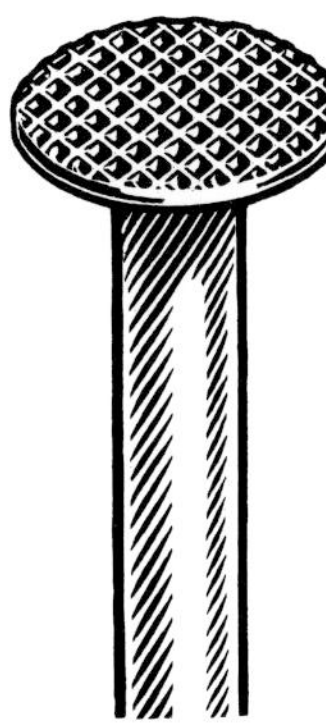
CHECKERED
Holds extra zinc or paint so hammering won't chip off the protective coating.

ROOFING
Oversize to reduce "tear-through" of roofing felt and asphalt shingles.

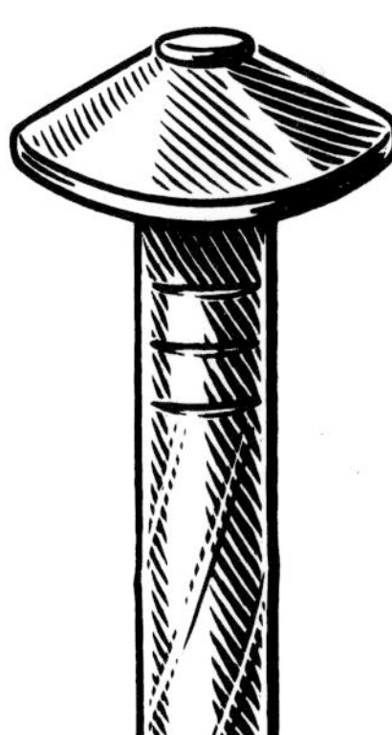
UMBRELLA
Wide rim allows nail to be driven without crushing corrugated metal roofing. Also helps cover the hole.

LEAD
For use on flat metal roofing. Soft lead dome, hammered flat, seals the hole.

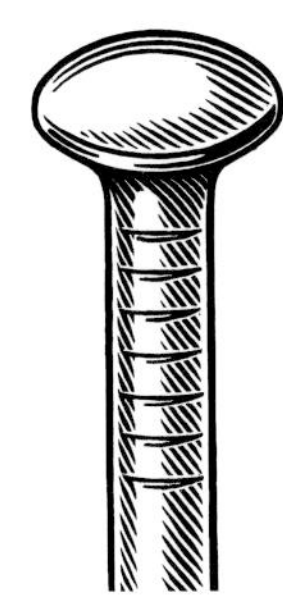
OVAL
Protects wood from hammer scarring. Found on old-style siding nails.

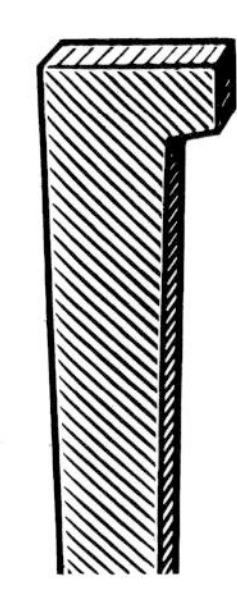
L-HEAD
Used on hardwood finish flooring. Those with two hooks are called T-heads.

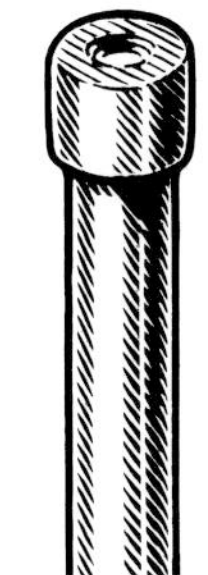
FINISH
Dimpled or cupped so it can be driven below the surface with a nailset, then puttied over. Heads on larger finish nails may not be dimpled.

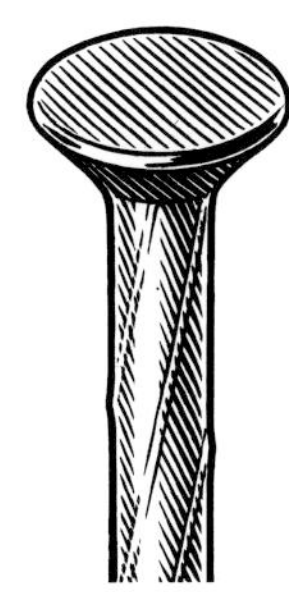
COUNTER-SUNK
Curved underside allows it to be driven flush without dimpling the wood.

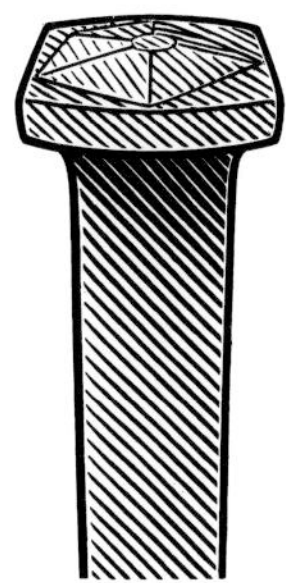
ROSEHEAD
On cut nails, raised area helps center hammer blows and provides decorative accent.

FINISH NAILS

Finish nails are among the lightweights of construction—and among the last driven—because their job is mainly to fasten the moldings and try to disappear. The vestigial heads are meant to be driven below the surface and puttied over. Tom Silva's tip: "If you try to drive them flush, you'll scar the wood. Drive them to head level, then sink them with a nailset." Casing nails are finish nails writ large. They're used for door and window frames and for exterior trim. Some are oil-hardened for use with hardwoods. Others are stainless or galvanized to inhibit rust. Bright or unfinished nails are for indoor use only. Smaller finish nails are called brads.

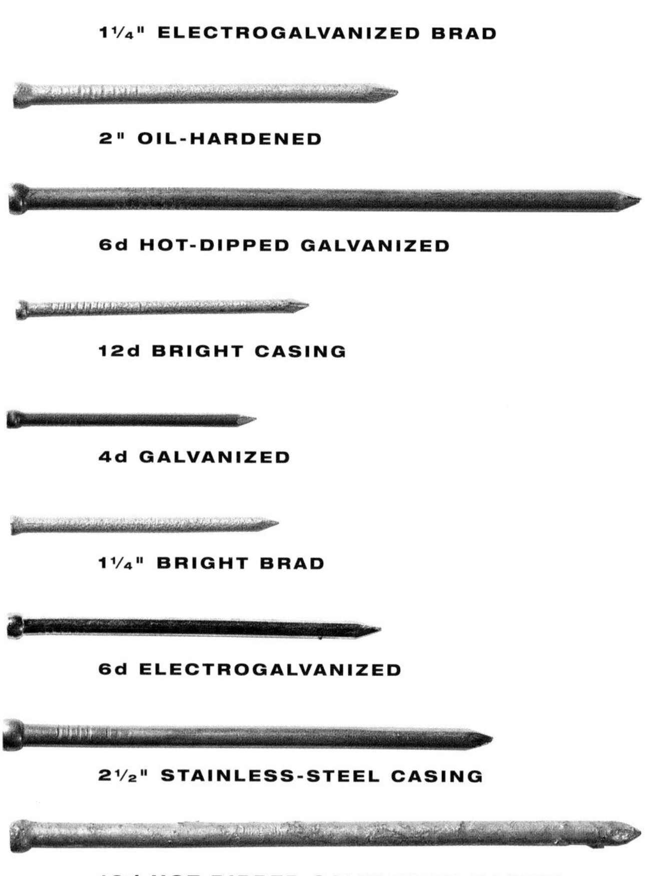

1¼" ELECTROGALVANIZED BRAD

2" OIL-HARDENED

6d HOT-DIPPED GALVANIZED

12d BRIGHT CASING

4d GALVANIZED

1¼" BRIGHT BRAD

6d ELECTROGALVANIZED

2½" STAINLESS-STEEL CASING

12d HOT-DIPPED GALVANIZED CASING

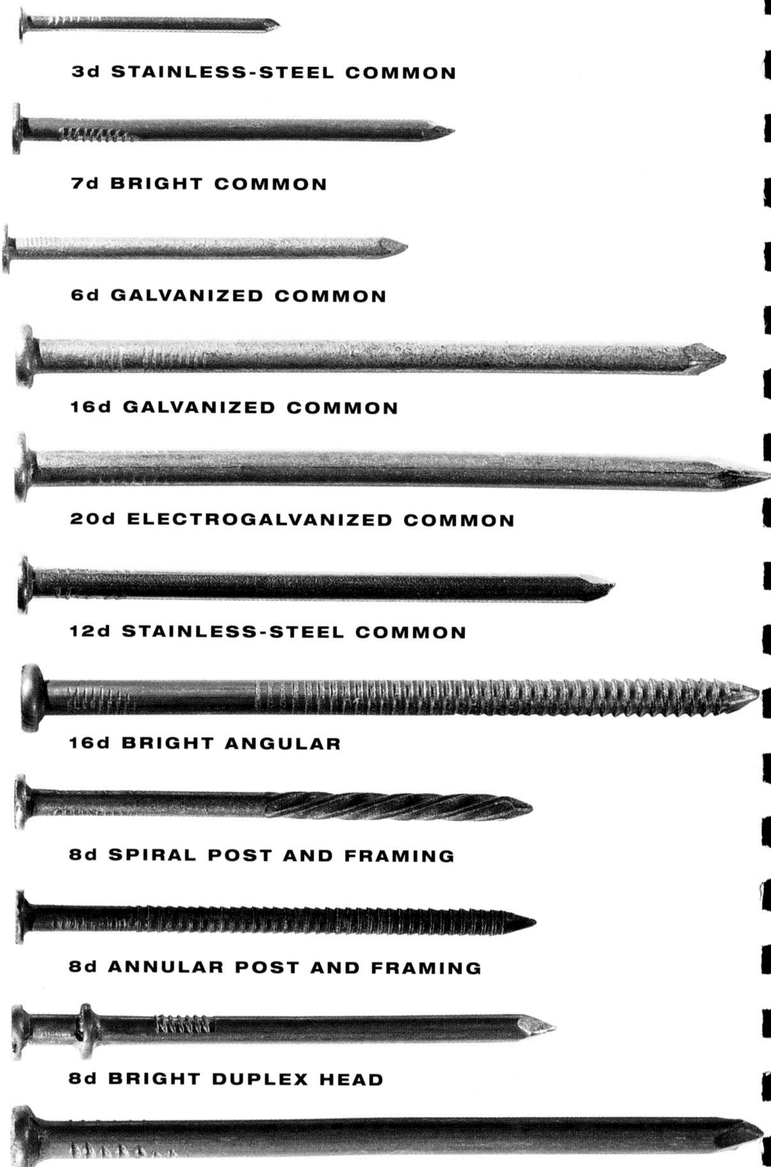

3d STAINLESS-STEEL COMMON

7d BRIGHT COMMON

6d GALVANIZED COMMON

16d GALVANIZED COMMON

20d ELECTROGALVANIZED COMMON

12d STAINLESS-STEEL COMMON

16d BRIGHT ANGULAR

8d SPIRAL POST AND FRAMING

8d ANNULAR POST AND FRAMING

8d BRIGHT DUPLEX HEAD

20d CEMENT-COATED SINKER

FRAMING NAILS

Most framers rely on common nails, which are thick-headed and thick-shanked to withstand blows from 20-, 22- or even 28-ounce hammers. They also use cement-coated sinkers (some believe they go in easier and hold better) and unfinished 8d (2½-inch) nails. Ring-shanks provide maximum holding power and can speed construction where building codes permit two of them to replace three common nails. Galvanized or stainless-steel nails are used when there's risk of corrosion. Easily removed duplex or double-headed nails are popular for framing scaffolding.

DECK NAILS

Heavy double spirals are used to fasten framing, while lighter ring- and spiral-shank nails lock down planking. Rustproof stainless, bronze or tempered aluminum outperform galvanized. Tom Silva says, "Side-nailing planks to metal deck clips means you need never see any nail heads. And your deck will last longer."

1½" SILICON-BRONZE ANNULAR

6d ALUMINUM SPIRAL

16d GALVANIZED SPIRAL

3⅝" GALVANIZED ANNULAR

4¼" STAINLESS-STEEL DOUBLE SPIRAL

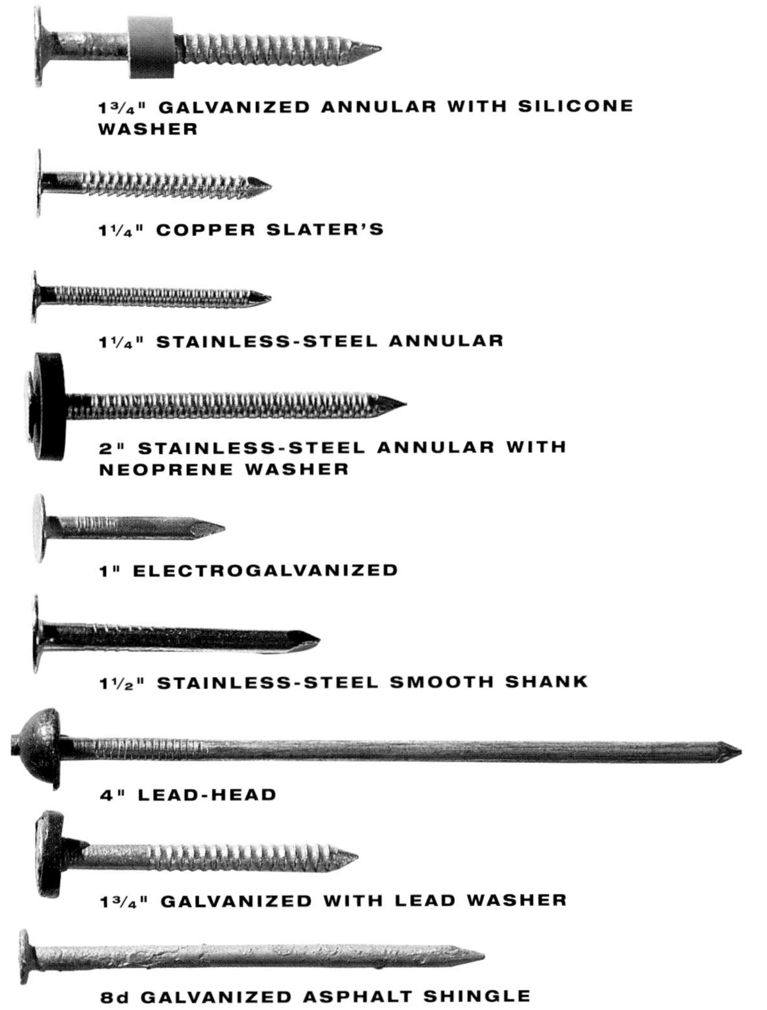

1¾" GALVANIZED ANNULAR WITH SILICONE WASHER

1¼" COPPER SLATER'S

1¼" STAINLESS-STEEL ANNULAR

2" STAINLESS-STEEL ANNULAR WITH NEOPRENE WASHER

1" ELECTROGALVANIZED

1½" STAINLESS-STEEL SMOOTH SHANK

4" LEAD-HEAD

1¾" GALVANIZED WITH LEAD WASHER

8d GALVANIZED ASPHALT SHINGLE

ROOFING NAILS

These vary widely in style and design, but if you're lucky you'll never see them again after they're driven. Galvanized asphalt-shingle nails are short on length and life span because asphalt shingles are thin and short-lived themselves. Among galvanized nails, double hot-dipped is the antirust standard. Stainless outlasts all competitors, but on long-lived tile or slate roofs, copper is preferred because it is virtually corrosion-proof and easier to cut when making repairs. Lead-heads are extra long for driving through the high ribs of corrugated metal roofing, and the lead helps seal the hole. So do the lead and rubber washers or cushions on nails used on flat metal roofs or flashing. Capped nails have oversize plastic or metal washers to hold down roofing felt.

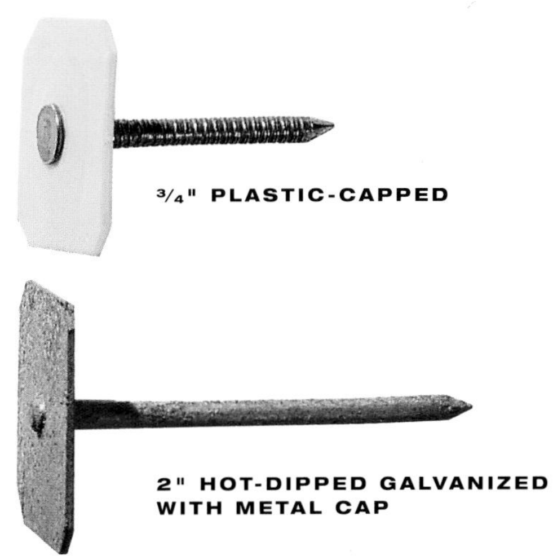

¾" PLASTIC-CAPPED

2" HOT-DIPPED GALVANIZED WITH METAL CAP

MASONRY NAILS

All masonry nails are made of hardened steel for driving into "green" concrete, which hasn't set rock-hard. (In cured concrete, pilot holes must be drilled first.) A 2-pound hand sledge is most often used. Masonry nails are brittle and can shatter, so wear goggles.

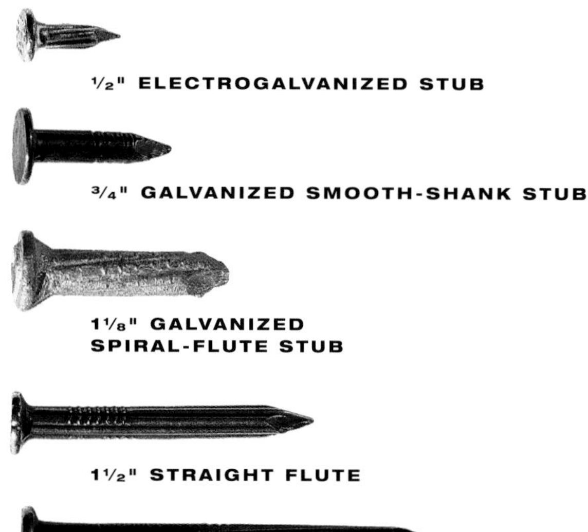

1/2" ELECTROGALVANIZED STUB

3/4" GALVANIZED SMOOTH-SHANK STUB

1 1/8" GALVANIZED SPIRAL-FLUTE STUB

1 1/2" STRAIGHT FLUTE

2" SPIRAL FLUTE

DRYWALL NAILS

Extra-sharp points speed installation, and ringed shanks resist popping. Some are cement-coated, a few are heat-treated to a slightly rust-resistant blue finish. Many are uncoated.

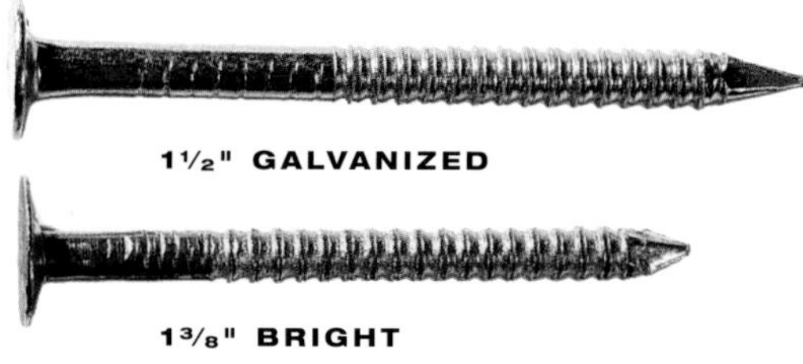

1 1/2" GALVANIZED

1 3/8" BRIGHT

SIDING NAILS

More like box nails than commons, they're slender and blunt to keep shingles and clapboards from splitting. They're also long and ring-shanked to resist plank cupping and warping. Stainless is the choice for redwood and cedar siding because it resists corrosion better than galvanized. Spiral nails are for hardboard siding. Capped or collared nails anchor foam sheathing.

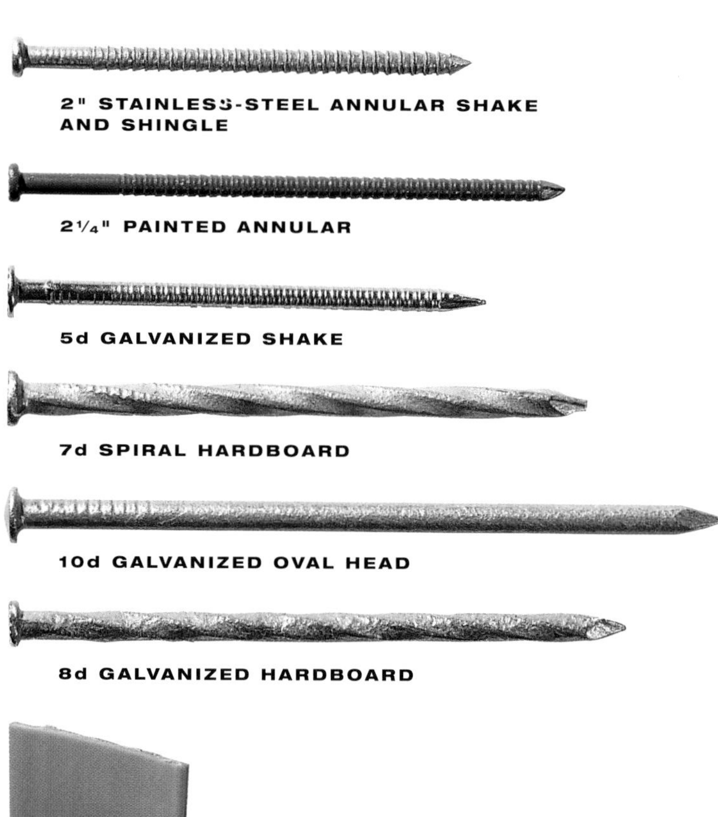

2" STAINLESS-STEEL ANNULAR SHAKE AND SHINGLE

2 1/4" PAINTED ANNULAR

5d GALVANIZED SHAKE

7d SPIRAL HARDBOARD

10d GALVANIZED OVAL HEAD

8d GALVANIZED HARDBOARD

1" PLASTIC-CAPPED

WROUGHT NAILS

Blacksmiths hammered out these oldest of nails, used to armor doors. Square-sectioned and usually square-headed, some have diamond heads or large flat panheads the size of quarters.

3 1/2" PYRAMID-HEAD DOOR

SOURCES

THE FOLLOWING COMPANIES SUPPLIED THE NAILS DEPICTED HERE

Clendenin Bros. Inc.
4309 Erdman Ave., Baltimore, MD 21213; 410-327-4500.

Dickson Weather-proof Nail Co.
Box 590, Evanston, IL 60204; 847-864-2060.

Georgia-Pacific Corp.
133 Peachtree St., Atlanta, GA 30303; 800-284-5347.

Manasquan Premium Fasteners
Box 669, Allenwood, NJ 08720; 800-542-1979.

Maze Nails
100 Church St., Peru, IL 61354; 815-223-8290.

Mazel & Co.
4300 W. Ferdinand St., Chicago, IL 60624; 312-533-1600.

Parker-Kalon Fasteners
510 River Rd., Shelton, CT 06484; 203-944-1711.

Parker Metal Corp.
Box 15052, Worcester, MA 01615; 800-225-9011 or 800-523-3002 (in MA).

Perma-Grip Fasteners
3375 N. Service Rd., Burlington, Ontario, Unit C-4, Canada L7N 3G2; 905-336-9400.

PrimeSource Inc.
1800 John Connally Dr., Carrollton, TX 75006; 214-417-3754.

Swan Secure Products Inc.
7525 Perryman Ct., Baltimore, MD 21226; 410-360-9100.

Tower Manufacturing Co.
1001 W. Second St., Madison, IN 47250; 812-265-4823.

Tremont Nail Co.
8 Elm St., Wareham, MA 02571; 508-295-0038.

Wheeling Corrugating Co. (La Belle)
134 Market St., Wheeling, WV 26003; 304-234-2400.

WROUGHT IRON NAILS BY SPECIAL ORDER

Woodbury Blacksmith and Forge Co.
161 Main St., Box 268, Woodbury, CT 06798; 203-263-5737.

Village Blacksmith Shop
221 North St., Goshen, CT 06756; 860-491-2371.

FLOORING NAILS

Ring-shanks pin down the underlayment, and spirals fasten the finish flooring when driven diagonally through the tongue (U.S. style) or the groove (French style). Cut nails are meant to be face-nailed through wide-plank country floors.

1¾" ANNULAR UNDERLAYMENT

2¼" OIL-HARDENED SPIRAL

8d CUT NAIL

BOX NAILS

These slender nails are used in cigar boxes, fruit crates and such because they don't split thin wooden slats. Antirust coatings allow wider light-duty use. And they are light: A pound of 8d common nails contains 106 pieces, but a pound of 8d box nails has 145.

4d BRIGHT

4d HOT-DIPPED GALVANIZED BOX

1¼" COPPER

⅞" WIRE

½" ELECTROGALVANIZED WIRE

SPIKES

These supernails, sometimes used by landscapers and timber framers, are 60d (6 inches) or longer. One monster is 135d (20¼ inches); it has a shank ⅞ inches square and weighs 4.4 pounds.

CUT NAILS

The late-18th-century development of nail-making machines ended the drudgery of hand-forging nails. Cut nails are easily recognized by their blunt points, flat sides and bias cuts that taper on two sides. They are most often used for restorations and wide-plank flooring. Rosehead stainless cut nails are also used for decorative effect on decks and siding. Their holding power is excellent until they're loosened—then they slide right out.

BRAD

BOAT

FLOORING

COMMON

ROSEHEAD CLINCH

WROUGHT HEAD

ROSEHEAD STAINLESS-STEEL DECK

PLASTIC NAILS

They're lightweight, they can't rust, they come in different colors and you can paint them. If you saw through one, it won't damage your blade. Plus, their tensile strength (resistance to withdrawal) is as much as 1½ times greater than that of smooth metal nails because they fuse with wood fibers. Is this the end of the venerable steel nail? Well, not yet. Raptor plastic nails' shear strength (resistance to snapping or bending) is about half that of a metal nail, so they're not suitable for framing. You can't hammer them; a special nail gun is required. And they won't work on hardwoods or medium-density fiberboard (known as MDF). But they might be just the thing for hanging siding, where shear loads aren't important but beating the elements is.

TOM SAYS

- To pick the right length: A nail should be two and a half times the thickness of the plank it goes through, but use common sense. If the point sticks out, shorten up.
- To start a nail: Don't grab it high. Hold it near the point, so your fingers are safe and the point is steadied on the wood. Tilt the head away from you a bit so the hammer hits it square.
- To avoid splitting wood: Use blunt-tipped nails or blunt what you have by whacking the points with your hammer. They'll ram through wood like a punch.

adhesives

POLYURETHANE GLUES

Everyone agrees that epoxy makes a strong waterproof bond, but it's tricky to use (its two components must be thoroughly mixed in the correct proportions), tough to sand and can trigger dermatitis. Yellow woodworking glue is safe and simple to use (there's no mixing), but its strength is so-so and you don't want it to get wet. Now there's a third alternative: so-called polyurethane glues (they're not true urethanes but isocyanates). These one-part, Belgian-made adhesives, marketed as Excel and Gorilla Glue, have nearly the strength of epoxy, are waterproof and can be used right from the bottle.

Polyurethane glues cure by reacting with moisture in the air; they reach full strength in 24 hours. Glue that squeezes out from joints foams up dramatically but is easily scraped or sanded and stained; there's no discernible glue line under the finish on the edge-glued mahogany pictured here. Precautions: Joints have to be tight; gaps larger than half a millimeter have impaired strength. The glue contains MDI, which can irritate the respiratory tract before it cures; good ventilation is recommended. (Asthmatics are isocyanate hypersensitive.) Also, wear gloves. Once this glue dries on your skin, you have to wait until it wears off. Polyurethane glues are expensive, on par with epoxy, but a little goes a long way, and there's less waste once you know how much glue a joint needs.

Raptor plastic nails, 2,000 DFS-8d per box, $60
Raptor Nail Division, Utility Composites Inc., Austin, TX 78759; 800-460-6933

Excel adhesive, #10036750, $18.65 for 750 ml
AmBel Corp., Box 819, Cottonport, LA 71327; 800-779-3935

Gorilla Glue, $19.95 for 18 oz
The Gorilla Group, Box 42532, Santa Barbara, CA 93140; 800-966-3458

Hook-and-loop Scotchmate fastening system, 1"x4.9 yds
3M Construction & Home Improvement Markets Div., 3M Center, Building 223-4S-01, St. Paul, MN 55144; 800-480-0611

Polyethylene tape, #NPT28, 2"x36 yds; $10
A. M. Leonard, Inc., 241 Fox Dr., Piqua, OH 45356-0816; 800-543-8955

Scotch metal repair duct tape, 1½"x10 yds; $7
3M.

Scotch Foam mounting double-sided tape, 1"x50" and ½"x75"; $2
3M.

Scotch self-bonding rubber splicing tape, ¾"x5 yds
3M.

Scotch indoor/outdoor carpet-seaming tape, 2½"x5 yds; $5-6
3M.

Fabric-backed cloth duct tape: industrial-grade #6B-28802, 2"x60 yds; in 7 colors; $7
Lab Safety Supply Inc., Box 1368, Janesville, WI 53546; 800-356-0783

Sheathing siding tape: #585CW2, 1⅞"x72 yds, $8
Venture Tape, 30 Commerce Rd., Box 384, Rockland, MA 02370; 800-343-1076

Painter's tape, Duo-Stick double-sided 1½"x36 yds; $8
Daubert Coated Products, Inc., 1 Westbrook Center, Westchester, IL 60154; 708-409-5100

Scotch vinyl electrical tape, ¾"x22 yds; $3
3M.

Scotch reinforced outdoor carpet tape, 1⅜"x13 yds; $6-7
3M.

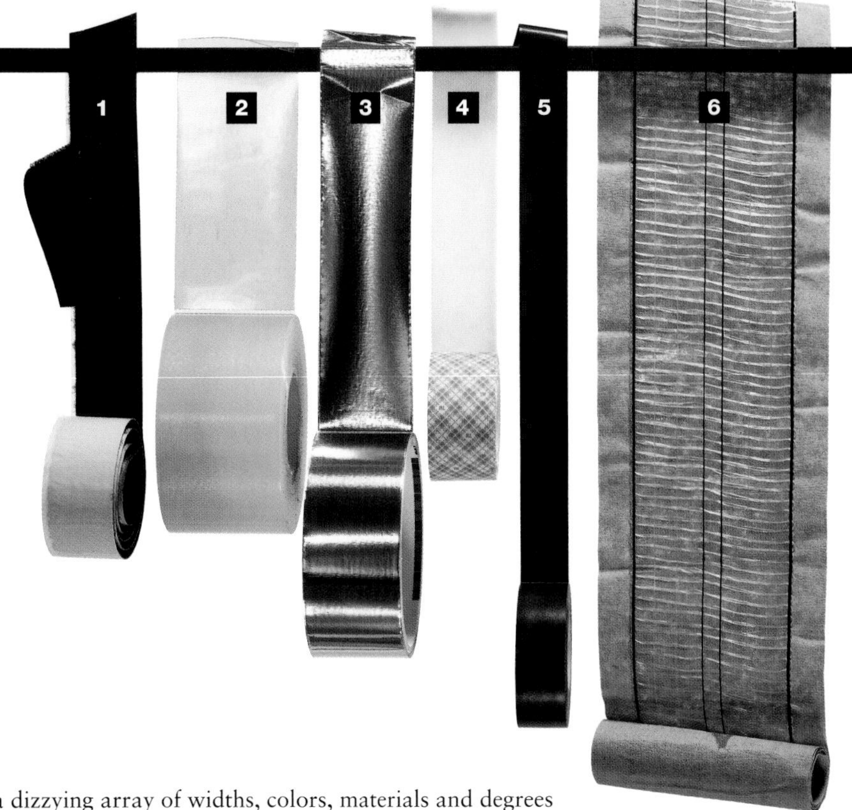

TAPES

Tape comes in a dizzying array of widths, colors, materials and degrees of stickiness. Which one to pick depends on the job: Some tapes need to be waterproof, others must be resistant to changes in temperature. Permanent repairs require strong adhesives, temporary uses need less holding power.

1. Hook-and-loop tape allows unlimited sticking and unsticking. **2. Polyethylene tape** stands up to heat, cold and damp; good for mending vinyl outdoor furniture. **3. Metal repair tape** follows contours easily, works well on gutters. **4. Foam mounting tape,** sticky on both sides, holds mirrors while mastic sets. **5. Self-bonding rubber tape** forms a waterproof mass; great in emergencies. **6. Fiberglass-reinforced indoor carpet tape** makes seams nearly invisible. **7. Fabric-backed duct tape** seals ducts, also works on broken tool handles. **8. Sheathing tape** seals joints in housewrap and insulation panels. **9. Painter's tape** holds drop cloths with an upper adhesive layer. **10. Vinyl tape** insulates electrical connections. **11. Patch-and-repair tape** mends torn bookbindings and ripped vinyl upholstery. **12. Outdoor carpet tape,** sticky on both sides, anchors carpet to a deck.

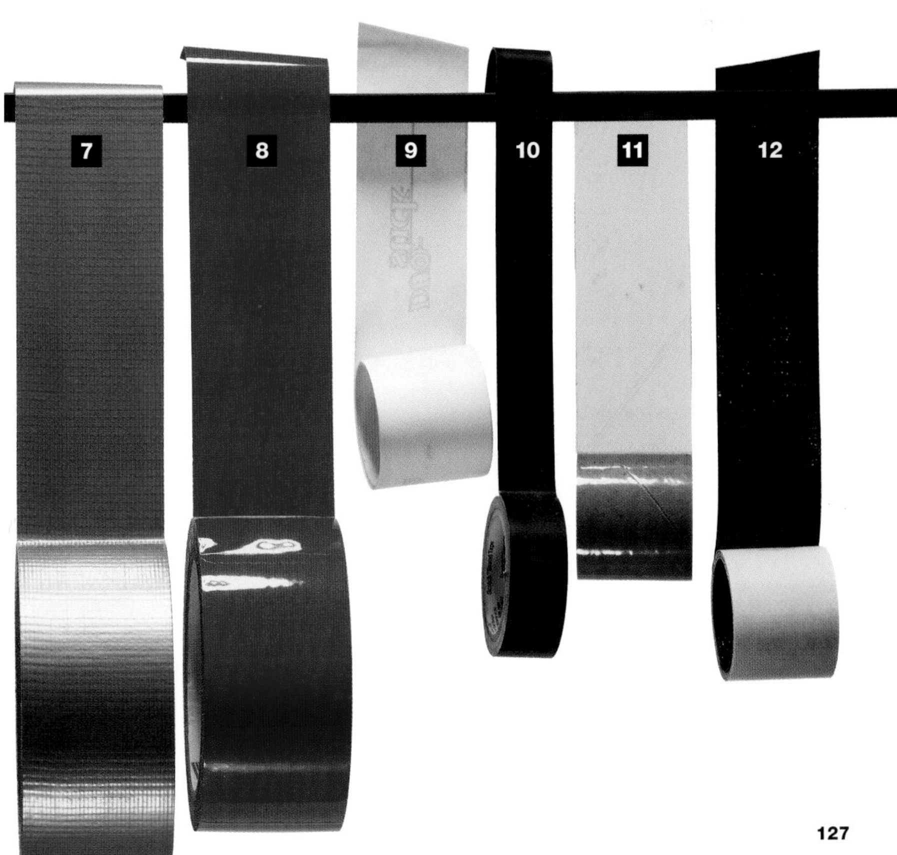

screws

A BRIEF HISTORY

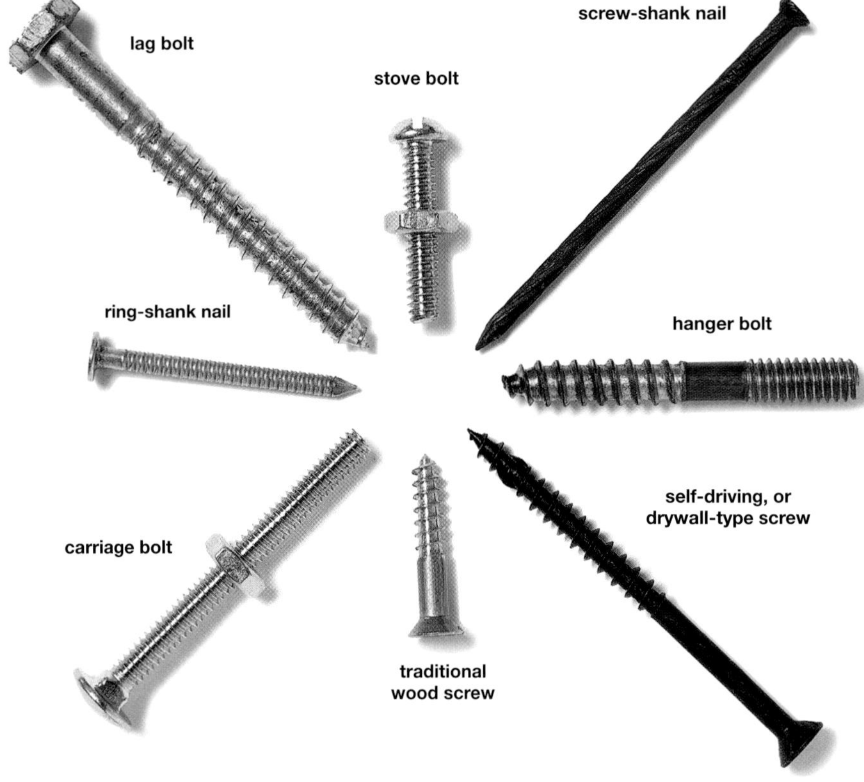

The current revolution in screws has its roots in the construction boom of the 1950s. Drywall hangers working on metal studs needed a quick, machine-driven method to hold gypsum panels in place. Their solution was the drywall screw—sharp and skinny, requiring no pilot hole and featuring a gently curved bugle head that could be countersunk without ripping the paper. Drywall hangers started using them on wood studs too, and manufacturers responded with coarser threads. Furniture makers discovered they also held well in particleboard and plywood. Then, in 1979, *Fine Woodworking* magazine ran a tip from a reader about how wonderful drywall screws were for all sorts of wood-holding tasks. Editors replied, "Wood screws are inferior; the only reason they continue to be used is ignorance."

The world of screws has never been the same. "That was the first time I ever saw anything written on this," says James Ray, president of McFeely's, a mail-order screw supplier. "After that point, everyone who wrote in with a tip started saying, 'Use a drywall screw.'"

Cordless drills (which can double as power drivers) became common, and at prices even hobby woodworkers could afford. People started using drywall screws where nails used to do. The screws held better, and there was no risk of a hammer blow jarring pieces out of alignment.

Other power-driven screw designs quickly followed. Many, like the original drywall screws, are self-driving, which means they push out a path by compressing the fibers in their way. Other new screws are truly self-drilling, which means they bore a hole through metal or wood, reducing strain on the screwdriver motor and keeping the material from splitting. Some screws can even cut threads in predrilled concrete, greatly simplifying a task that used to require big holes fitted with plastic or lead anchors. The slotted head rarely appears on these new screws because drivers used with them slip easily. At minimum, the screws have Phillips heads. Norm Abram predicts that most screws soon will have square recesses, which virtually eliminate "cam-out," the annoying tendency of a screwdriver to lift up and out of the slot.

5 BASIC SCREWS

Screws of the new generation are straight from head to tip, never tapered like traditional wood screws. Called rolled-thread screws, they are formed by pressing indentations into a wire blank instead of cutting away excess metal on a lathe. This results in a stronger screw, because the grain of the metal is uninterrupted. Rolled-thread screws are not as new as they may seem: The first patent was issued in 1836, but the technology did not catch on until better alloys were developed.

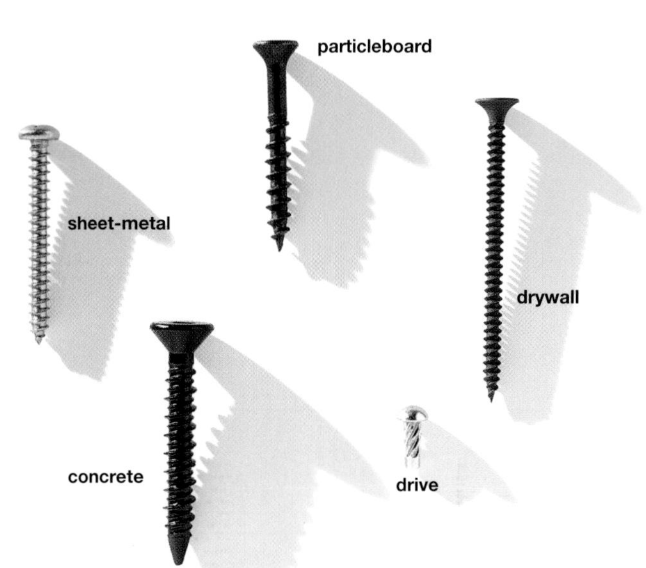

JUST WHAT IS A SCREW?

The American National Standards Institute defines the screw as an externally threaded fastener. Its definition of a bolt is identical. But in common usage, screws are relatively short (less than 4 inches), pointed and threaded into the materials they fasten. Bolts are long, blunt fasteners that thread into nuts. Bridging the two categories are machine screws (small and blunt) and lag bolts (big and pointed). Then there are drive screws, more commonly called screw-shank nails. With helically threaded shanks, they qualify as screws. But they are driven with hammers—precisely how most people define nails. Ring-shank nails, though they look similar, are definitely not screws. Their cone-shaped ridges form a series of rings rather than a continuous thread up the shank. Beyond defining parts of a screw and insisting that they be no longer than labeled, the standards-setting arm of U.S. industry is silent about what screws must be made of. "No standard says anything about material requirements," says Steve Winistorfer, a research engineer at the Forest Products Research Laboratory in Madison, Wisconsin. Screws are made and sold, he explains, generally without any warranty of performance or means of tracing them once installed.

DRYWALL SCREWS: NOT ALWAYS A SAFE CHOICE

James Mortensen, a contractor in Charlton, Massachusetts, spent three months on his back after a roofing bracket secured with 3-inch drywall screws collapsed. The screws sheared off all at once, teaching Mortensen what metal experts have long known: Drywall screws, though wonderfully suited to their initial task, are hardened in a way that makes them unsafe for carrying heavy shear (lateral) loads. The screws have no "fatigue zone," where they might stretch out of shape but not break; they just snap. "Assume they have zero-rated strength," says Joseph R. Loferski, associate professor of wood science and forest products at Virginia Tech in Blacksburg, Virginia. Drywall screws should never be used to carry structural loads. Nor should they be used to hang kitchen cabinets or heavy bookshelves. What about deck screws that look like drywall screws? They're probably fine for decks. But assume they are brittle too. "We have a whole division that imports drywall screws by the container load, coats them and sells them for deck screws," one executive told us. Drywall-shaped screws made of stainless steel in the 300 series are safe for shear loads; they're not brittle.

THE RIGHT MATERIAL FOR THE JOB

Choose screws made of a material suited to the task. Steel, tough and cheap, is most common but rusts if wet. Though a zinc coating helps, galvanized screws can still stain redwood and cedar and corrode in pressure-treated wood. Some patented finishes cover zinc with baked-on paint to prevent galvanic reactions, which can occur whenever two dissimilar metals are exposed to water. Check labels carefully: Finishes not backed up by a zinc layer can scrape off while the screw is turning, leaving the metal free to rust with the first rain. Norm Abram's choice for outdoor use is stainless steel. It is safe with all woods and has no coating to chip off.

ALUMINUM: little strength; for light jobs only.

SILICON BRONZE: doesn't rust, costly; boaters' choice.

SOLID BRASS: very soft; mostly for decorative uses.

MECHANICALLY GALVANIZED: inexpensive; zinc is pounded on.

YELLOW-ZINC PLATED: more protection than bare zinc.

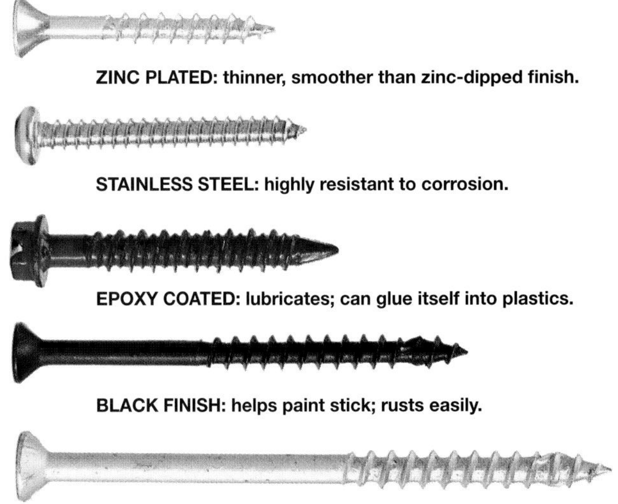

ZINC PLATED: thinner, smoother than zinc-dipped finish.

STAINLESS STEEL: highly resistant to corrosion.

EPOXY COATED: lubricates; can glue itself into plastics.

BLACK FINISH: helps paint stick; rusts easily.

DUROCOAT: multi-layer; lubricates, resists corrosion.

A GUIDE TO SCREWS

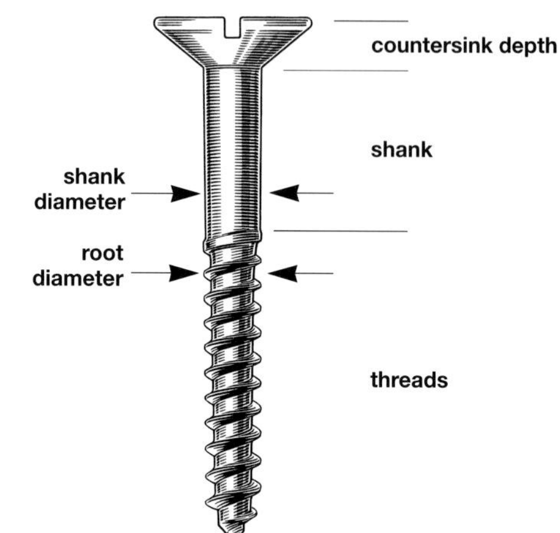

Screws come in a seemingly infinite variety of styles engineered to accomplish specific tasks. Which features are best for a particular application? Here's a guide to decoding new and traditional screws.

A screw's length specifies how far in it will drive. Countersinking screws are measured from top to tip; flush screws, from back of head to tip. Length is given in inches. Screw diameter indicates maximum width, generally at edge of threads, and it is specified in gauge numbers. No. 0 is 1/16-inch thick. Each additional gauge adds .013 (about 1/64) of an inch.

HEADS

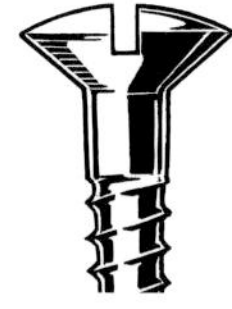

OVAL

Used where appearance counts. Back bevel allows most of head to be countersunk.

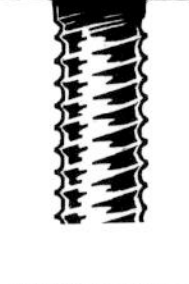

ROUND

Sits flush on surface. Used where counter-sinking is not practical or desired.

PAN

Wide head allows great clamping strength. Not very attractive, so usually used where not seen.

TRUSS

Similar to pan head, but lower profile makes it popular for furniture.

INTEGRAL WASHER

Wide head holds well even if shank hole is oversize to allow for adjustment.

BINDING

Thicker head than pan allows deeper slot. Often used in electrical work for good contact.

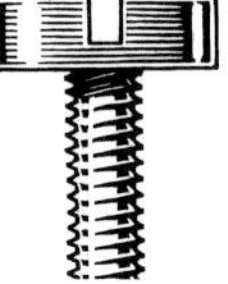

CHEESE

Thick head allows deeper slot for increased driving power.

FILLISTER

Advantages of cheese head, but with slightly rounded top for better appearance.

FLAT

Designed to be countersunk so top sits flush. For neat job, countersink must be predrilled.

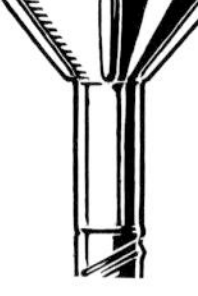

FLAT WITH NIBS

Ribs on back of head bore countersink, even in difficult materials like hardwoods.

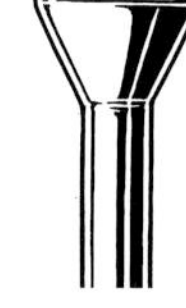

TRIM HEAD

Similar to finish nail but bigger. Used where large head would look ugly.

BUGLE

Gentle curve on back allows head to sink itself into soft materials.

WAFER

Similar to bugle head, but wider and thicker. Good for soft materials.

HEXAGONAL

Wrench or driver grips entire head, allowing great torque. This one has built-in washer.

BREAK-OFF

Tamper resistant. Hex head is tightened with socket wrench, then snapped off.

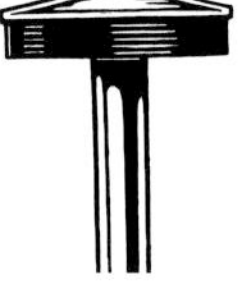

SELF-SEALING

Built-in metal washer backed by neoprene layer seals against leaks.

THREADS

When choosing a thread, pay most attention to what works best in the underlying material, not what's being fastened to it. The goal is to maximize thread contact but minimize effort needed to turn the screw.

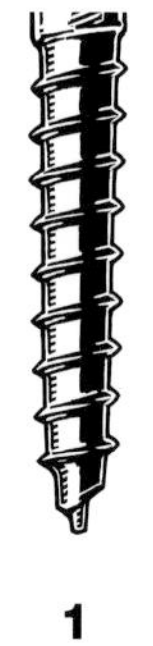

1

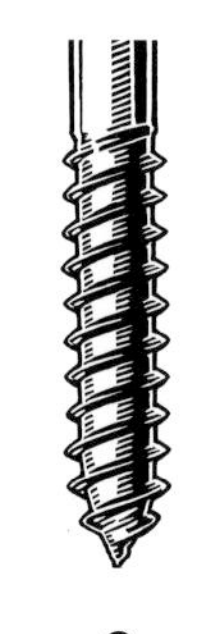

2

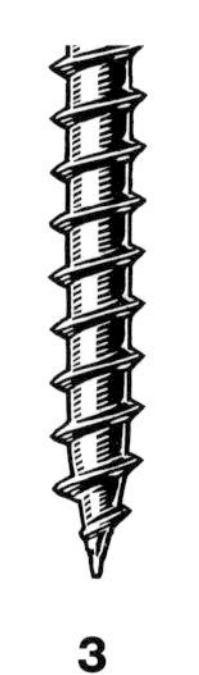

3

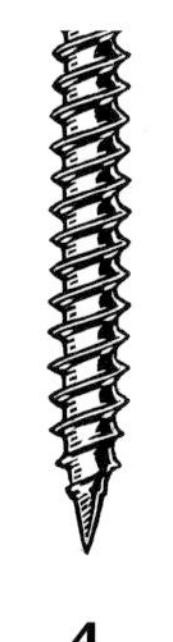

4

5

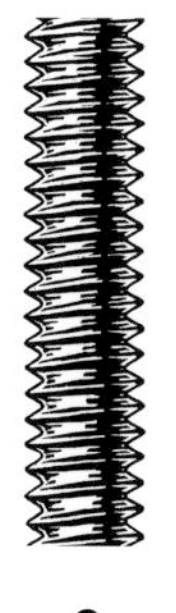

6

7

1. Cut thread: These relatively shallow threads, found on traditional wood screws, are cut on a lathe. They work well only if proper pilot holes have been drilled. **2. Rolled thread (fine):** These threads are pressed into the blank, so they extend beyond the shank. No-waste manufacturing method actually creates stronger screws because metal fibers aren't cut. Sharp, closely spaced threads are especially good for attaching things to thin metal because they allow more threads to be in contact. Also useful in hardwoods. **3. Coarse thread:** Deep, widely spaced threads are especially good for holding mushy fibers such as those in softwood and particleboard. Wide spacing allows screw to drive in fast. This example has a single thread, or lead. **4. Double lead:** Parallel threads work up the shank. These give the holding power of tightly spaced threads but require only half as many turns to be sunk home. For wood or metal, depending on depth and spacing of thread. **5. Hi-Lo thread:** In this patented double-thread design, one thread is deeper than the other for easier driving and better holding. Common on concrete screws. **6. Machine screw:** Fine, closely spaced threads hold better than other options for fastening into metal. But machine screws require a tapped hole (or a nut) because they cannot form mating threads. **7. Drive screw:** Steep angle of threads allows fast insertion, usually with a hammer. Some have slots for easy removal; smooth-headed versions can only be drilled out.

DRIVES

Every screw needs a way to be twisted in and, usually, out. Sometimes a wrench fits over the entire head. More often, the head has a recess into which a screwdriver can be fitted. Slotted heads came first but work worst because flat screwdrivers slip easily. Phillips drives were an improvement but still allow slippage. (In fact, they were designed to prevent overtightening of screws on aluminum aircraft.) Other designs give better control, sometimes in more ways than one: Tamper-resistant screws can't be removed without special screwdrivers that are often difficult to find.

SLOTTED

Useful now mostly on antiques, where other options would look out of place.

PHILLIPS

Cross-drive recess has tapered, flat-bottomed slots. Named for 1935 inventor, Henry M. Phillips.

FREARSON

Lesser-known cross-drive, with less tapered slots. Also called Reed & Prince and Type II.

SQUARE

Also known by the name of its 1908 inventor, P.L. Robertson. Norm Abram's favorite drive.

COMBINATION

Accepts screwdrivers with either Phillips or square tips.

TORX

Sunburst recess has no taper. Popular in industry because drivers don't slip.

CLUTCH

Bow-tie recess is mostly found on mobile homes and in electric motors.

HEX RECESS

Found on headed and headless screws. Headless setscrews are adjusted with Allen wrenches.

HEXAGONAL

Wrench or driver grips entire head, allowing great torque.

ONE-WAY

Tamper-resistant. Can be installed with regular screwdriver but removed only by special tool.

PIN-IN-HEAD

Tamper-resistant design looks similar to Phillips drive, but pin blocks all but special tool.

DRILLED SPANNER

Tamper-resistant. Can be used with magnetic drivers that grip the whole head.

POINTS

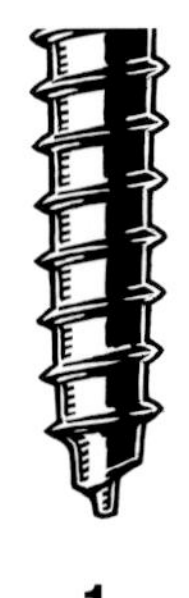

1

2

3

4

5

6

7

1. Gimlet point: This point, traditional for wood screws, helps align screw with pilot hole. Screw can be forced into softwood without a pilot hole but with great risk of splitting the wood. **2. Self-driving:** Sharp tip on skinny screw usually drives into thin metal or softwood without a pilot hole. Tip doesn't actually cut, however, so a pilot hole is still needed near edges and in hardwood or thick metal. The most common tip for power-driven screws. **3. Fastap:** Patented self-drilling design for wood has sharp tip to keep screw from wandering. Flutes ream out pilot hole to virtually eliminate splitting, even in hardwoods or near ends of boards. **4. Auger point:** Another self-drilling design; has sharp tooth cut into the first several threads so that screw cuts its own pilot hole and pushes away waste. **5. Self-drilling, for metal:** This tip and the threads that follow drill a pilot hole and form mating threads in metal. Various lengths of drilling flutes are available, depending on the thickness of the metal. **6. Winged Teks:** This patented tip is designed to attach wood to metal studs. The wings just above the tip bore a relatively wide pilot hole through the wood, then break off when they hit the metal so that the hole in the stud is sized properly. **7. Serrated teeth:** Teeth machined into lower threads help prevent screw from backing out, even after prolonged vibration. Useful with plywood, fiberglass, laminated wood veneers and composite materials.

HOW TO DRESS UP A SCREW

One way to hide flat-head screws is to counterbore deeply, drive the screw, then glue in a wooden plug. Options shown here are more suited to self-driving screws; no drilling or countersinking is needed.

Patented Snap Caps, near right, are held in place by a washer through which the screw is driven. Similar products snap onto square-recess screws. At far right, a metal trim washer surrounds the screw.

A SCREW'S HOLDING POWER

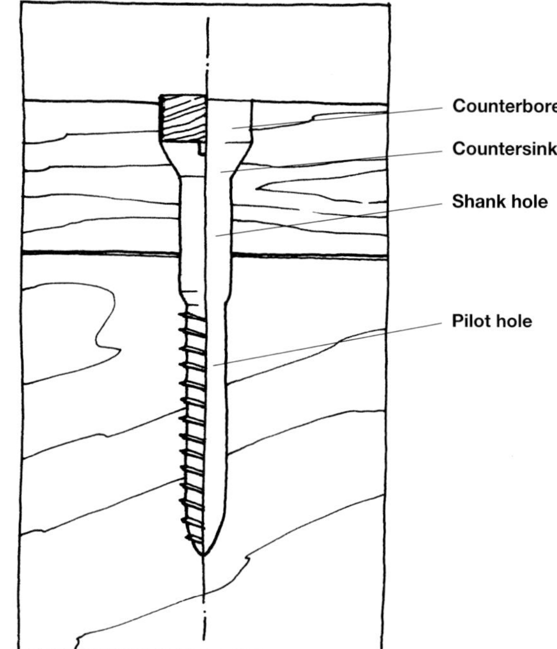

Think of screws as miniature clamps instead of nails with ridges and you'll avoid many mistakes. A nail works by wedging itself between wood fibers. Once the nail is hammered in, the fibers spring back toward their original position, applying pressure that keeps the nail in place. A screw works differently. None of its holding power comes from threads that may be embedded in the piece being fastened. The important action takes place in the material it's being fastened to. Threads dig in and pull the screw head until the two pieces are tight. The underside of the head is one end of the clamp; the threads sunk in the back piece are the other. Any threads that engage in the top piece can actually interfere with the clamping action, which is why correctly sized pilot holes are so important. Because threads in the back piece are so crucial, screws should be long enough so that two-thirds of their length will be there. For more holding power, try longer screws before thicker ones, because holding power increases more with length than with diameter. Nails hold best when driven at a slant in opposing directions. But no one would set up clamps at a slant. Screws should be driven in straight.

Installing a traditional wood screw properly can mean drilling up to four holes of different sizes: a shank hole in the top piece, no narrower than the screw shank; a tapered pilot hole in the bottom piece, 60 to 70 percent of the root diameter in softwood and 80 to 90 percent in hardwood; a countersink to accept the underside of the head; and a counterbore to hold a wooden plug over the screw head, if desired. Newer screw designs often eliminate the need for pilot holes. When predrilling is needed (near board ends, in hardwoods and where pieces can't be held firmly together), a single, straight hole fits both shank and root.

TIP

Any screw drives more easily if first scraped on a cake of wax. Don't use soap; it can absorb water from the wood, causing rust.

Screws should pass freely through the top piece.

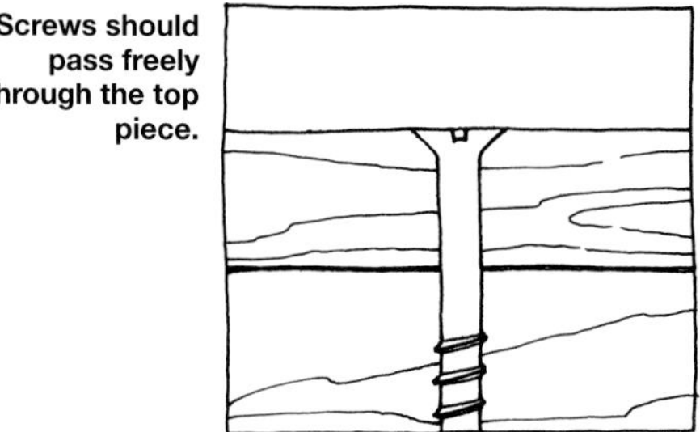

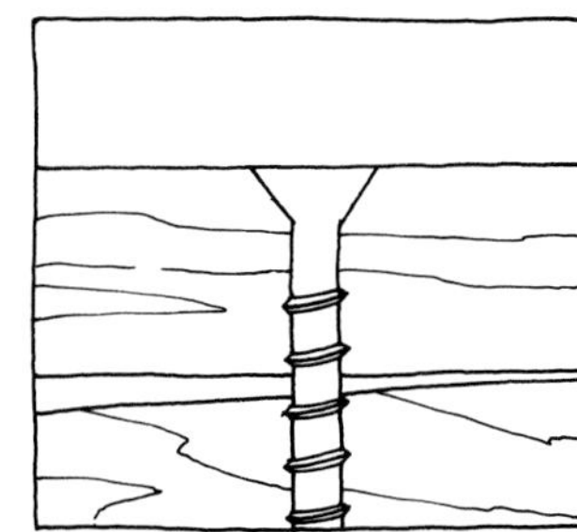

Threads in the top layer can keep the joint from being pulled tight. If this occurs, back out the screw, hold the pieces tight and try a screw with a longer shank. Prevent problems by predrilling or clamping.

SQUARE-DRIVE SCREWS

Square-drive screws, which get their name from the square, tapered recess in the head right where you'd expect to find a slot or a Phillips star, have been around since 1908. Properly called Robertson screws, after Canadian inventor P. L. Robertson, they were developed to overcome "cam out" (when a screw-drive bit loses its grip), and are now more available. Norm recommends them for woodworking projects: "They feel more positive and have less slippage. I think they could dominate the industry in 15 years."

SELF-DRILLING SCREWS

Fastap self-drilling screws are one of the niftiest little inventions to come our way in years. Anyone who has used drywall screws in endless ways that have nothing to do with drywall knows they have a nasty habit of snapping off. Fastaps don't do that because the tiny cutting flutes at the tip of the screw drill out a path for the screw that follows. Nibs on the underside of the head make the screws self-countersinking too. They range from 1¼ inches to 6 inches long for exterior use, and 1 inch to 3 inches for interiors. Both square-drive and Phillips-head versions are available.

McFeely's Square Drive Screws
1620 Wythe Road, Box 3, Lynchburg, VA 24505-0003; 800-443-7937.

Stillwater Fasteners
Box 128, East Freetown, MA 02717; 508-763-8044.

Elco Industries Inc.
1111 Samuelson Road, Box 7009, Rockford, IL 61125-7009; 815-397-5155.

Fastap self-drilling screws: 3" exterior screws with Durocoat, $7 per lb.
Faspac Inc. 13909 N.W. 3rd Court, Vancouver, WA 98685; 800-847-4714.

ANCHOR PRICES LISTED BY PIECE UNLESS OTHERWISE SPECIFIED

Plastic expansion anchor: #8, 4 cents; #10, 5 cents
Star Expansion Co.

Nylon nail-ins: ¼"x1", $19 for 100
Powers (Rawl) Products.

Legs expanding wallboard anchor: $24 for 100
Powers (Rawl) Products.

Scru'N'Grip threaded anchors: plastic, 28 cents; metal, 30 cents
Star Expansion Co.

Drive anchor: ½", 61 cents; ⅝", 66 cents
Star Expansion Co.

Four-legged anchor: ½", 39 cents; ⅝", 51 cents
Star Expansion Co.

Toggle anchor: ¼"x3", 75 cents
EZ-T Fastner Co. 3292 S. Bannock St., Englewood, CO 80110; 800-854-3279.

E-Z Toggle: 79 cents
ITW Buildex 1349 W. Bryn Mawr Ave., Itasca, IL 60143; 800-323-0720.

Plastic toggle anchor: ½", 25 cents; ⅝", 28 cents
Star Expansion Co. Route 32, Mountainville, NY 10953; 800-247-8274.

Rawl toggle bolt: ¼"x3", $50 for 100
Powers (Rawl) Products, 2 F.B. Powers Square New Rochelle, NY 10802; 800-524-3244.

anchors

DRYWALL ANCHORS

Little more than a paper-and-gypsum sandwich, wallboard often must bear hefty loads—bookshelves, towel bars and artwork, to name a few. Drilling through wallboard into a stud is always best, but when a good stud is hard to find, try a drywall anchor. These come in three types: toggle anchors that clamp against the drywall, screw anchors with deep threads that embed themselves in the wall and expansion anchors that expand as a screw or nail is driven in. Manufacturers rate their anchors for shear (pulling sideways) and tensile (pulling out) loads. Don't look for the ratings on the package; they're seldom included. Call the manufacturer or check the catalog instead. Anchors should be able to hold four times the weight of the object being hung, but remember that even the best anchor can't exceed the strength of the drywall. When mounting, don't place anchors too close—no less than 10 anchor diameters apart for expansion types. And measure carefully before sinking the anchor. Adjustments after the fact are almost impossible, and the hangers leave gaping holes in the wall.

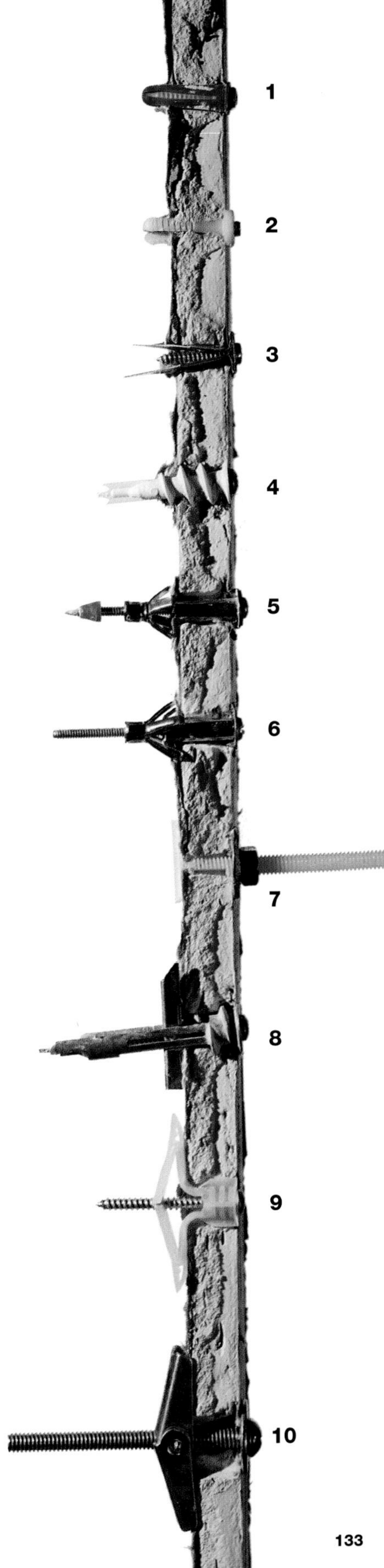

1. A **plastic expansion anchor** leaves only a small hole in the wall when it's removed, but a wrong-size screw or pilot hole will render one of these useless.
2. **Nylon nail-in anchors** provide more holding power than nails alone but can't handle heavy-duty loads. They are more commonly used on masonry. Of all the anchors we surveyed, these first two types offer the least holding power.
3. This **expanding anchor** is hammered in, leaving only a slit in the wallboard if it's ever removed. A lot of torque is required to drive the screw and spread the metal wings.
4. **Threaded anchors** are quick to install, require no predrilling and can be removed and used again. Their holding power ranks below the toggle bolts but above expanding anchors.
5. **Drive anchors** are similar to four-legged anchors but are hammered, not screwed, into place. Predrilling is still advisable since punching a hole through the wallboard makes the core crumble, compromising the wall's strength.
6. This **four-legged toggle anchor** splays as the screw is tightened but works only on the wallboard thickness it's rated for. Check instructions for the approximate number of turns required to set the anchor: Overtightening damages and weakens the drywall.
7. This **toggle anchor** does the least damage to the wall, requiring a hole just large enough to insert a slender nylon stem.
8. We like this **toggle** because it's one of the simplest anchors to install (no predrilling required). It's also among the strongest. The toggle opens automatically when the screw is inserted.
9. This **plastic toggle anchor** tucks into a point and fits through a small pilot hole, then opens like a flower. But if the drywall is too thick, the toggle won't open at all, so its holding power is nil.
10. **Toggle-bolt anchors** are the workhorses, able to hold the heaviest tensile loads. They also leave a big scar in the wall and require great talent or an extra pair of hands to install. The spring-loaded wings are inserted first through a hole up to three times the diameter of the screw. Tightening the screw is awkward because the toggle must be pulled taut against the board while turning. If the screw is removed, the toggle drops behind the wall, gone forever.

notes

salvaged and recycled materials

REUSING MATERIALS | ARCHITECTURAL SALVAGE

RECYCLED GLASS TILES | SIMULATED GRANITE

RECLAIMED LUMBER | OLD WINDOW GLASS

reusing materials

Choose to Reuse, by Nikki and David Goldbeck, is a fat book (480 pages) listing suppliers of used materials across the country. It tipped us off to Urban Ore in Berkeley, which has been so successful selling items destined for the dump that it's gearing up to offer franchises. With an inventory that produced revenues of $1.4 million last year, it's certain to have something you can use. We picked up a leaded glass window, a (new) brass faucet, a baluster and a post support, all at prices well below retail.

If you have a computer and modem, try the National Materials Exchange Network, a free bulletin-board service. The emphasis is on industrial by-products (want a few vats of acid?), but wood, metal and other construction materials are there too. Organizer Bob Smee said he sees great potential for exchanges of building materials, especially since listings can be sorted by area code. Call 509-466-1532 for information. The World Wide Web address is http://www.earthcycle.com/nmen. Modem users without Internet access can call the bulletin board at 509-466-1019.

RESTORING AND REUSING AN OLD FAUCET

BEFORE

This solid brass 1920s American Standard faucet was found jumbled in a drawer at The Brass Knob, a Washington, D.C., salvage yard. Some of its parts were missing, others damaged. The nickel plating was almost all worn off. One of the porcelain cross handles was cracked, and the porcelain escutcheons didn't fit properly—they originally belonged to the matching shower set.

THE COST

Restoring an old faucet is generally cheaper than buying a comparable reproduction, but this process is not for bargain hunters. The true reward is saving a beautiful piece of work. Restored, this faucet cost $275; if you had found the fixture and done some of the work at home, the cost would have been closer to $150. A new nickel-plated, porcelain-detailed faucet would cost $300 to $500.

ANTIQUE FAUCET: $75

REBUILDING: $125

PARTS $75
LABOR $50

REPLATING: $75

TOTAL COST: $275

AFTER

The restored faucet was installed in a new porcelain pedestal sink with an 8-inch spread (the distance between taps). Building codes in some parts of the country would mandate backflow preventers on the feed lines of this faucet because the spout is less than 1 inch above the basin rim. Check with the building inspector before installing any antique fixture.

Choose to Reuse, $18.95
Ceres Press, Box 87, Woodstock, NY 12498; 914-679-5573.

Urban Ore
1333 Sixth St., Berkeley, CA 94702; 510-559-4460.

The Brass Knob
1701A Kalorama Rd. NW, Washington, DC 20009; 202-986-1506.

Adkins Architectural Antiques
3515 Fannin St., Houston, TX 77004; 713-522-6547.

Architectural Antiques
801 Washington Ave. North, Minneapolis, MN 55401; 612-332-8344.

The Bank
1824 Felicity St., New Orleans, LA 70113; 504-523-2702.

Decorum Hardware
231 Commercial St., Portland, ME 04101; 207-775-3346.

Florida Victorian Architectural Antiques
112 W. Georgia Ave., Deland, FL 32720; 904-734-9300.

Irreplaceable Artifacts
14 Second Ave., New York, NY 10003; 212-777-2900.

Ohmega Salvage
2407 San Pablo Ave., Berkeley, CA 94702; 510-843-7368.

Olde Theatre Architectural Salvage
2045 Broadway, Kansas City, MO 64108; 816-283-3740.

Pinch of the Past Architectural Antiques
109 W. Broughton St., Savannah, GA 31401; 912-232-5563.

The Salamander and the Web, a finding service for architectural salvage; $38 membership includes monthly newsletter
Box 1834, Topeka, KS 66601-1834; 800-984-3932.

Salvage One
1524 S. Sangamon St., Chicago, IL 60608; 312-725-8243.

United House Wrecking
535 Hope St., Stamford, CT 06906; 203-348-5371.

architectural salvage

Columns and cornices, fanlights and newel posts: Craftsmen's details are grace notes of style. Look carefully and you'll find pieces of the past, from leaded glass windows to claw-footed tubs, being sold by salvage yards across the country. Antique shops, too, often have isolated bits and pieces of old houses, as do commercial demolition companies, which may salvage parts from their wrecking jobs. With time on your hands you can nose about promising-looking renovation sites as a save-it-yourselfer, but the most reliable sources of architectural salvage are specialists. Serious salvage yards don't wait for things to turn up serendipitously—they're constantly on the prowl and have their own workers to handle the dismantling.

Salvage means rescue, but it also means saving money and time. Period renovations often require items that are no longer made and expensive to reproduce. Victorian brass faucets may have to be sacrificed for lack of a valve stem, and special-order windows may take weeks to arrive.

INVESTIGATE BUILDING CODES

New building codes can be hostile to old fixtures and fittings. Victorian toilets are charming but wasteful; federal law requires newly installed toilets to use just 1.6 gallons of water per flush. Faucets can be trouble if spouts clear the sink by less than an inch (2 inches for tubs). And those handsome etched-glass door panels? In some situations, entry doors are required by local building codes to contain shatterproof glass. So always review your municipality's codes before you write the check.

SALVAGE YOUR OWN

Contractors often accept spot cash for recyclable items; keep your eyes peeled for upcoming demolitions (dumpsters are always promising signs). If not allowed to remove stuff yourself (usually due to insurance risk), make the deal contingent on undamaged removal with the contractor's help. And check community "big trash" days, when hard-to-jettison items are set out. We got a turn-of-the-century pedestal sink free that way; it would have cost $300 or more from a dealer.

recycled glass tiles

With a movement afoot to build clean and green, many architects and contractors now use recycled materials as a matter of course. This helps small, low-tech businesses like Bedrock Industries, which turns trash from recyclers, restaurants and tempering plants into treasure. Whether it starts out as a brilliant blue water bottle or crystalline shards of tempered glass, each tile emerges with a particular texture and hue, as displayed below.

Cobalt blue tile is made from Arizona Iced Tea or Welsh Tynant water bottles. Though the bottles look nearly identical, glass from each is fired separately.

Wine bottles can be used to fire a batch of evergreen-colored tiles.

When fired, clear juice bottles create opaque tile, often used for countertops and floors.

Chardonnay bottles give this "celery" tile its hue.

Irregularly surfaced clear tile is made from tempered glass. A favorite for use in bathrooms, clear tiles are usually bedded in a white Thinset fortified with acrylic. Glass tiles are treated like granite, cut with a wet saw and a diamond blade.

Recycled glass tiles: $19 per sq. ft.
Bedrock Industries, 620 North 85th St., Seattle, WA 98103; 206-781-7025; distributed exclusively by Pratt & Larson Tile Inc., 1201 SE Third Ave., Portland, OR 97214; 503-231-9464; and 207 Second Ave. South, Seattle, WA 98104; 206-343-7907.

Tessera hand-cut mosaic: $24 per sq. ft.
Oceanside Glasstile Co., 3235 Tyler St., Carlsbad, CA 92008; 619-434-0051.

Oceanside tiles: Oceania series, 3"x3" blue field, $1.20 each; 6"x6" blue diamond spiral, $9.45 each; 4"x4" rainbow X tile, $5.25 each
distributed exclusively by Walker & Zanger, 8901 Bradley Ave., Sun Valley, CA 91352; 818-504-0235.

Toltec tile: 6"x6" spiral, $14.95 each
Toltec Architectural Glass, 3901 W. Jefferson Blvd., Los Angeles, CA 90016; 213-732-7654; distributed by Walker & Zanger.

Environ simulated granite: 3"x6" sheet, ⅛" thick, $54.36; 3"x6" sheet, 1" thick, $116
Phenix Biocomposites Inc., Box 609, Mankato, MN 56002-0609; 800-324-8187.

Clockwise from lower left: Hand-cut mosaic, rainbow X tile, blue diamond spiral, Toltec spiral and blue field tile, all made from 85 percent recycled material. The tile molds are formed in clay, then cast in iron and filled with molten glass. The iridescent colors are created with a metallic solution bonded to the front of the tile at 2,000 degrees Fahrenheit. Such tiles are often used decoratively with other varieties of tile and slate or to line swimming pools.

APPLICATIONS

Left: After two years of hard wear on the floor of Seattle's Pratt & Larson tile showroom, there isn't a scratch. "It's a labor of love to make something like this, and you can see that in the tile," says co-owner Katherine Blakeney, who sells Bedrock tiles for about $19 a square foot. Middle: One client used the tiles to line a barbecue area overlooking Lake Washington, in colors to match the blue of its waters. Right: Architect Robert Harrison installed a sauna and shower using clear tiles. "The owner was after the notion of a luminous cave," he says.

simulated granite

Recycled newspapers and soy flour seem unlikely ingredients for a building product, but Phenix Biocomposites, a Minnesota-based company, has found a way to combine the two to yield a smooth, hard board called Environ that is suitable for anything from tabletops to panel inserts in cabinets. Soon to come: wall tiles and moldings. Six of the 10 colors take on the look of granite with just a few coats of clear finish. That's because the manufacturers add bits of color to the newspapers, giving the composite a speckly grain that simulates stone. Woodworkers report that the mixture of glue and newsprint cuts easily, with no fuzz or chipping.

reclaimed lumber

Would you look for lumber at the bottom of a river? If you needed choice wood for flooring or a cabinet front, would you don scuba gear? That's what George and Carol Goodwin and their crew do. Goodwin Heart Pine of Micanopy, Florida, makes a business of recovering century-old logs from riverbeds, drying and milling them and selling the prized lumber.

Heart pine is the wood of the longleaf pine, Pinus palustris. These 80- to 125-foot-tall trees used to cover 40 percent of the southeastern coastal plain from Virginia to eastern Texas. Once the most common of woods, it is now so rare and valuable that it's worth the time and effort needed to reclaim it. Salvagers remove it from old buildings or fish it by hand out of the muddy bottom of the Suwannee River, as the Goodwins do.

BUYING AND USING RECLAIMED LUMBER

Goodwin Heart Pine sells three main grades of lumber: select, vertical and curly. The terms are confusing because they encompass both sawing technique and presence of knots. Customers who order select will get wood with a few tight, sound knots. Only if they specify FAS (First and Second grade) will even one surface be free of knots. Regardless, select boards will be flatsawn (or plainsawn), simply sliced from the log. Viewed on end, flatsawn boards have annual growth lines running nearly up and down at the edges and horizontally in the middle. Thus, they can cup as the growth rings straighten out. Not so with vertical sawn boards, which account for about 25 percent of Goodwin's sales. Their growth rings run nearly perpendicular to the face, so the wood is more stable.

Most lumberyards charge about the same for a 12-inch-wide board as for three 4-inch-wide boards of the same thickness and length. Because its supply is limited, Goodwin charges a steep premium for wide, thick boards. Regular select lumber costs $3.95 a board foot for pieces 1 inch thick and 3 inches wide, but $11.58 for the same volume in pieces 2½ inches thick and 12 inches wide. Prices for select and vertical grades don't vary according to whether wood is rough, planed smooth or milled for tongue-and-groove. And the rare curly grade? It's $16 or more a board foot, with no choice as to knots.

LUMBER IDENTIFICATION

Recycled barns, flooring from old homes, beams from industrial buildings erected earlier this century, river-recovered cypress and Douglas fir: There's a world of reclaimed wood on the market.

If you have your hands on some salvaged lumber but don't know what kind of wood you've got, mail a piece of it to the U.S. Forest Service's research lab in Madison, Wisconsin, and experts there will identify it for you. The service is free to anyone (with a limit of five samples a year). Just send a chunk of the wood, along with a letter saying what you know about it. "Send in a reasonable-size piece," says Regis Miller, who supervises the program. "What we really like is 6 inches off a board. That doesn't mean we can't work with less. But if I say less, man, I really get a lot less."

SOURCES

River-recovered heart pine lumber
Goodwin Heart Pine Co., Rt. 2, Box 119-AA, Micanopy, FL 32667; 904-466-0339; fax 904-466-0608.

Heart pine source and installer
Authentic Pine Floors, Inc., 4042 Highway 42, Box 206, Locust Grove, GA 30248; 800-283-6038.

FOR MORE INFORMATION

Photocopies of out-of-print books and a 32-page calendar ($8) about longleaf pine
Association for the Restoration of Longleaf Pine, Box 141464, Gainsville, FL 32614-1464.

WOOD IDENTIFICATION SERVICE

Free. Send sample and a letter stating where the wood came from to
Center for Wood Anatomy Research, Forest Products Laboratory, USDA Forest Service, 1 Gifford Pinchot Dr., Madison, WI 53705-2398.

Old glass: Light Restoration Glass, 10"x14", $14.95 per sq. ft.
S.A. Bendheim Co. Inc., 61 Willett St., Passaic, NJ 07055; 800-221-7379.

FOR FURTHER READING

***Repairing Old and Historic Windows*, New York Landmarks Conservancy, 1992, $24.95**
John Wiley & Sons, 1 Wiley Dr., Sommerset, NJ 08875; 800-225-5945.

***The Window Handbook*, Charles E. Fisher III, ed., 1986, $32**
Historic Education Foundation, Box 77160, Washington, DC 20013; 202-828-0096.

***The Window Workbook*, 1986, $48.25**
Historic Preservation Education Foundation.

"The Repair of Historic Wooden Windows," John H. Meyers, 1981, Preservation Brief #9
Preservation Assistance Division, National Park Service, Box 37127, Suite 200, Washington, DC 20013; 202-343-9573.

Above left: Heart pine is prime flooring material because it wears well. River-recovered logs are especially prized because they can be cut into wide boards like these 11-inch pieces. Goodwin does not center the tongue-and-groove on its flooring; this is so the thicker "half" can face up, allowing more resurfacings.
Above right: In its time, heart pine was a utility wood, used mostly for framing, floors and boats. But now that it's rare, its attractive grain patterns and strength give it a place in furniture, cabinets and paneling. Goodwin offers a line of stair parts, including single-piece and glued-up stair treads and blanks for balusters, rails and molding.

CURLY HEART PINE

One in every 200 longleaf pine trees develops this distinctive grain; instead of lines, it runs in whorls and curls. A pattern of bumps on the log's surface tells the salvager he has found a prized curly-grain log, worth about four times as much as the others. It will most likely be sold for cabinet doors or fine furniture.

old window glass

The curved striations in this old pane are the telltale signature of crown glass.

Like an old, handcrafted wooden window sash, old glass—with its wavy modulations and other imperfections—is a valuable artifact in its own right, one that deserves to be preserved.

Until the early part of the 20th century, all the glass for windows was made by hand—and mouth—through the prodigious efforts of artisans skilled in the manipulation of this brittle material. From the early 17th century right up to the mid-1800s, windows were made with crown glass. The glassmaker spun a ball of molten glass on the end of a pontil rod until it formed into a disc, or crown, about 3 to 5 feet in diameter.

Mouth-blown cylinder glass, long an inferior alternative to crown glass, became the window glass of choice in the mid-19th century, when improved techniques enabled panes to be made bigger, faster and more cheaply.

Compressed air and machinery replaced lungs and hand-craftsmanship in 1905, when factories began making machine-blown cylinder glass, using a technique that created towering glass cylinders up to 40 feet tall and 2 feet in diameter. Cylinder glass began to be replaced in the 1920s by drawn glass, produced from sheets instead of cylinders, and plate glass, a poured and polished glass made primarily for automobiles. In 1958, float glass—so called because the glass sheets are flowed out on a bed of molten tin—introduced a hitherto impossible distortion-free and defect-free uniformity. Now, with virtually all window glass being manufactured with the float-glass method, the distinctive flaws of old glass are almost irreplaceable. When you consider what it took to produce old glass, casually throwing it away seems a shame.

One way to find old glass is to salvage it from windows that are being tossed out. You may also find old window glass at landfills or on the street during bulk trash pickup days. To achieve the same antique look, you can buy reproduction "old glass," but at $15 per pane, it's as expensive as it is beautiful.

for more information

BOOKS

Complete Book of Kitchen Design, Ellen Rand and Florence Perchuk, 1991, 216 pp., $17; Consumer Reports Books, 101 Truman Ave., Yonkers, NY 10703; 914-378-2000.

The Complete Guide to Sharpening, Leonard Lee, 1995, 256 pp., $22.95; The Taunton Press, 63 S. Main St., Box 5507, Newtown, CT 06470-5506; 800-888-8286

Installing and Repairing Plumbing Fixtures, Peter Hemp, 1994, 184 pp., $19.95; The Taunton Press, 63 S. Main St., Box 5506, Newtown, CT 06470-5506; 800-888-8286.

Measure Twice, Cut Once: Lessons from a Master Carpenter, Norm Abram, 1996; 196 pp., $17.95; Little, Brown & Co., 34 Beacon St., Boston, MA 02108; 800-759-0190.

The Old House Journal Guide to Restoration, Patricia Poore, editor, 1992, 400 pp., $39.95; Dutton/Penguin Books USA, 120 Woodbine St., Bergenfield, NJ 07621; 800-253-6476.

This Old House Kitchens: A Guide to Design and Renovation, Steve Thomas and Philip Langdon, 1992, 273 pp., $24.95; Little, Brown & Co., 34 Beacon St., Boston, MA 02108; 800-759-0190.

This Old House Bathrooms: A Guide to Design and Renovation, Steve Thomas and Philip Langdon, 1993, 270 pp., $24.95; Little, Brown & Co., 34 Beacon St., Boston, MA 02108; 800-759-0190.

The Timber Frame Home: Design, Construction, and Finishing, Tedd Benson, 1988, 225 pp., $24.95; The Taunton Press, 63 S. Main St., Box 5507, Newtown, CT 06470-5506; 800-888-8286.

Twentieth Century Building Materials: History and Conservation, edited by Thomas C. Jester, National Park Service 1996, 352 pp., $55; McGraw Hill Co., 11 West 19th St., New York, NY 10020; 800-722-4726.

Understanding Wood Finishing, Bob Flexnor, 1994, 320 pp., $27.95; Rodale Press, 33 East Monir St., Emmaus, PA 18098; 800-848-4735.

Fixing and Avoiding Woodworking Mistakes, Sandor Nagyszalanczy, 1995, $20; The Taunton Press, 63 South Main St., Box 5507, Newtown, CT 06470-5507; 800-888-8268.

PAMPHLETS

"An Illustrated Guide to Hardwood Lumber Grades," No. G-102, 1994 Edition; $5, National Hardwood Lumber Association, Box 34518, Memphis, TN 38184-0518; 901-377-1818.

"Imagination Within: Idea Book for Architectural Interiors," No. M-504, $4; National Hardwood Lumber Association.

"Low Flow Toilets," Article #9994, $7.75; Consumer Reports; 800-766-9988.

"The Preservation and Repair of Historic Clay Tile Roofs," A.E. Grimmer and P.K. Williams, Preservation Briefs No. 30, U.S. Dept. of Interior, National Park Service, Box 37127, Washington, DC 20013; 202-208-7394.

"Wood Flooring: A Lifetime of Beauty," National Wood Flooring Association, 233 Old Meramec Station Rd., Manchester, MO 63021; 800-422-4556.

PERIODICALS

Adhesives and Sealants Industry magazine, Box 400, Flossmoor, IL 60422; 708-922-0761

Hardwood Floors magazine, 1846 Hoffman St., Madison, WI 53704; 608-249-0186.

Journal of Light Construction, RR No. 2, Box 146, Richmond, VT 05477; 800-552-1951.

Old House Journal and *Old House Interiors,* 2 Main St., Gloucester, MA 01930; 508-283-3200.

Preservation, the magazine of the National Trust for Historic Preservation, 1785 Massachusetts Ave. NW, Washington, DC 20036; 202-673-4000.

ASSOCIATIONS

American Plywood Association—The Engineered Wood Association, Box 11700, Tacoma, WA 98411; 206-565-6600.

American Wood Preservers Institute, 1945 Old Gallows Rd., Suite 550, Vienna, VA 22182; 703-893-4005.

Association for Preservation Technology International, APT Box 8178, Fredricksburg, VA 22404; 703-373-1621.

National Hardwood Lumber Association, Box 34518, Memphis, TN 38184-0518; 901-377-1818.

National Institute of Building Sciences, 1201 L St. NW, Suite 400, Washington, DC 20005; 202-289-7800; fax 202-289-1092.

National Wood Flooring Association, 233 Old Meramec Station Rd., Manchester, MO 63021; 800-422-4556.

Sealant, Waterproofing & Restoration Institute, 3101 Broadway, Suite 585, Kansas City, MO 64111; 816-561-8230.

Society for the Preservation of New England Antiquities, Harrison Gray Otis House, 141 Cambridge St., Boston, MA 02114; 617-227-3956.

The Gypsum Association, 810 First St. NE, Washington, DC 20002; 202-289-5440.

The Timber Framers Guild of North America, Box 1075, Bellingham, WA 98227; 360-733-4001.

Western Wood Products Association, Yeon Building, 522 SW Fifth Ave., Portland, OR 97204; 503-224-3930.

Western Red Cedar Lumber Association, 1100-555 Burrard St., Vancouver, BC V7X 1S7; 604-684-0266.

Western Wood Preservers Institute, 601 Main St., Suite 405, Vancouver, WA 98660; 206-693-9958.

credits

Many people contributed to the issues of *This Old House* magazine from which most of this book is drawn.

WRITERS

Thomas Baker, Don Best, Claudia Glenn Dowling, Peter Edmonston, Paul Engstrom, Mark Feirer, Jeanne Huber, Brad Lemley, Peter Lemos, Ben Lloyd, William Marsano, Stephen Petranek, John S. Saladyga, William Sampson, Wendy Talarico, Charles Wardell

PHOTOGRAPHERS

Greg Anthon, David Barry, Dan Borris, Neil Brown, Craig Cutler, Davies & Starr, Carlton Davis, Furnald/Gray, Michael Grimm, Darrin Haddad, Lynn Johnson, Spencer Jones, Keller & Keller, Marco Lavrisha, Barry David Marcus, Joshua McHugh, Micheal P. McLaughlin, Martin Mistretta, Daniel Moss, J. Michael Myers, Douglas Rosa, Rosa & Rosa, Aldo Rossi, Chris Sanders, James Schnepf, Kolin Smith, Reinaldo Smoleanschi, Wayne Sorce, Luca Trovato, William Vazquez, Simon Watson, Kevin Wilkes, James Wojcik, James Worrell; additional photographs courtesy of The Tile Man Inc. and Western Red Cedar Lumber Association

ILLUSTRATORS

Tungwai Chau, Wilton Duckworth, Stan Fellows, Clancy Gibson

SPECIAL THANKS TO

Norm Abram, Steve Thomas, Tom Silva, Richard Trethewey, Russ Morash and Bruce Irving at the show; Karen Johnson and Peter McGhee at WGBH; Isolde Motley at the magazine